Handbook of
PLANT TYPES

Handbook of
PLANT TYPES

Sandra Holmes

Hodder & Stoughton

A MEMBER OF THE HODDER HEADLINE GROUP

British Library Cataloguing in Publication Data

Holmes, Sandra
  Handbook of plant types.
  1. Botany—Classification
  I. Title
  581'.012   QK95

ISBN 0 340 39106 5

First published 1987
Impression number 10 9 8 7 6 5 4
Year   1999 1998 1997 1996 1995
Copyright © 1987 Sandra Holmes

Phototypeset by: BH Typesetters & Designers, Wardington Manor, Wardington, Nr. Banbury, Oxfordshire.

Printed in Great Britain for Hodder & Stoughton Educational, a division of Hodder Headline PLC, 338 Euston Road, London NW1 3BH by Athenæum Press Ltd, Gateshead, Tyne & Wear.

Contents

Preface

The purpose of this book is to provide a guide to the classification and structure of the plant types that appear on most advanced level biology and botany syllabuses, and in other courses of this level.

The plant types are illustrated by means of annotated diagrams, each with a classification of the plant and notes on its habitat, reproduction, and other details which vary from plant to plant. Drawings are of a simple line type, showing features of biological interest, in the style required by the examining boards, and that could be reproduced by students. Plant classification includes the main group (division or phylum), the class, and one or two other subgroups. The plant may be illustrated by a generalised diagram of the genus, or of one particular species. The plant divisions considered are the Algae, Fungi, Lichenes, Bryophyta, Pteridophyta, and Spermatophyta. Life cycles are given where appropriate, and each flowering plant is shown as the half flower, floral diagram, floral formula, and fruit, with notes on pollination and fruit dispersal. The flowers are followed by sections on seeds and vegetative propagation. Bacteria are not considered because they cannot be seen clearly with the optical microscope. Scale is not given on the diagrams since it is expected that students will study actual specimens, and because some plants show considerable variation in size.

The plants chosen are those with which urban students will be familiar. The flowers illustrated are common and in bloom in early summer, or are those with a fruit which students would be expected to have studied. The moss shown is the very common wall moss, *Tortula*, not the rarer *Funaria* of the textbooks, and the mushroom described is the cultivated, not the field species. Seeds and seedlings are discussed in relation to man with, for example, the parts of the wheat grain used in various types of bread, and notes on the differences in the amino acids stored in cereals and legumes. There are sections on the economic importance of algae and fungi, and a list of food plants on our supermarket shelves, each with its Latin name, family, and the nature of the part eaten.

Classification has been kept as simple as possible with regard to A-level requirements, but accounts of the divisions include groups relevant to a study of evolution (such as *Rhynia*, *Ginkgo* and the cycads) and to ecology (such as the characteristics of all ten classes of algae), as well as the plants required for a study of the diversity of plant life.

Since syllabuses vary, each annotated diagram and life cycle can be read more or less in isolation without sifting through unwanted material, but emphasis is placed on evolutionary relationships and the concept of alternation of generations in the life cycles of plants. In the classification sections, characteristics are given in note form; for all levels belows that of a division, characteristics are listed as numbered points and can be made into comparison tables by students. Botanical terminology is defined in the glossary. Where there are alternative names for a structure, one term is used consistently in the text and the alternatives are included in the glossary.

It is intended that this book will fill the gap between the modern comprehensive functional approach type of textbook and the older structural botanical tomes which provide far more detail than is required in today's syllabuses.

Introduction: plant classification

The main categories used in plant classification are:

Kingdom
Division (or phylum)
Class
Order
Family
Genus (plural genera)
Species

The term subdivision is sometimes used between division and class, especially in seed plants, and the term subclass is used between class and order.

A species has two names, the genus (generic) name and the specific epithet. This is known as a binomial or binominal. The naming of species is called binomial nomenclature, and follows a system worked out by the Swedish botanist Linnaeus in the eighteenth century, so it is also called the Linnaean system. The names of species and genera are printed in italics and underlined in manuscript. All other categories are printed in Roman type and are single words, known as uninomials.

There are certain rules and conventions governing the naming (nomenclature) of plants. These rules are codified in the International Code of Botanical Nomenclature (I.C.B.N.) which is updated every few years. The I.C.B.N. has laid down rules for the endings (suffixes) of the categories, so that the level of a group can easily be recognised. For example, orders end in -ales, such as Agaricales, Filicales, and families end in -aceae, such as Ranunculaceae, Liliaceae. From the order level downwards, these suffixes are the same for all plants, but above the order level they are different for algae, fungi, and all other plants. The suffixes are:

	Algae	Fungi	Other plants
Division	-phyta	-mycota	-phyta
Subdivision	-phytina	-mycotina	-phytina
Class	-phyceae	-mycetes	-opsida
Subclass	-phycidae	-mycetidae	-idae
Order	-ales	-ales	-ales
Family	-aceae	-aceae	-aceae

Although most families end in -aceae, there are nine exceptions in flowering plants, sanctioned by long usage. The Compositae, Gramineae, Labiatae, Leguminosae and Palmae are some of these exceptions.

The above system has only been introduced quite recently and was finalised in the 1970s. There are two main problems with using it; firstly that old and varied terminology is still frequently used, and secondly that there is disagreement among experts about the status of plant groups. For example, if the Algae is considered to be a division, it should be renamed Phycophyta; green algae are then a class called Chlorophyceae, brown algae are Phaeophyceae, etc. But some experts think that Algae should be given an indeterminate status above a division but below a kingdom, and the classes should be raised to division status as Chlorophyta, Phaeophyta, etc. The same problem arises with the Fungi. In some systems the Fungi are considered to form a division, sometimes called Mycophyta, while in other systems the Fungi are given a status above that of a division, sometimes even that of a kingdom, and the slime moulds and true fungi are then given division status as Myxomycota and Eumycota.

Among the other plants, these new suffixes make well known groups difficult to recognise. For example, the mosses, formerly known as Musci, have been renamed Bryopsida, and the dicotyledons should now be called Magnoliopsida. Ferns are properly called Filicopsida or Polypodiopsida, or sometimes Pteropsida, but the old name of Filicineae is still seen. The term Pteropsida is confusing because it is sometimes used for a group which includes ferns, gymnosperms and angiosperms on the grounds that they are closely related. Angiosperms are properly called Magnoliophytina and Gymnosperms are Coniferophytina.

In the text, an intermediate course is taken between the correct new system and the old familiar one. The terms Angiosperms, Gymnosperms, Dicotyledons and Monocotyledons are used as proper classification terms, but Bryopsida and Hepaticopsida are used for mosses and liverworts. In the fungi, subgroups commonly used are given subclass status but the proper subclass suffix is not used. This is because students are more likely to meet the terms Discomycetes, Pyrenomycetes, etc., than Discomycetidae, Pyrenomycetidae, especially since the status of the groups vary, and the former terms are sometimes used as common names.

The formation of common names from the Latin is complex and somewhat arbitrary. We are familiar with it in everyday life. The genus *Antirrhinum* is printed in italics and with a capital letter, but the term is also used as a common name, antirrhinum. Both forms are correct and their use depends on context. The same is true of rhododendron, delphinium and many other plant names. It is also true, with a little more subtlety for algae, fungi, angiosperms, gymnosperms, monocotyledons and dicotyledons. When discussing "the division Algae" a capital letter is used, but when discussing "freshwater algae" it is not. Most names both lose their capital letter and change their ending when used as common names, as in the case of the division Bryophyta which becomes bryophytes, and a member of the Ascomycetes which becomes an ascomycete, and so on. In general, when a term is used with its category name, a capital letter and the correct suffix are found; when a term is used as a common name, a lower case first letter and an anglicised ending are used. Both are seen in the text here, as occasion demands. Students can be confused about this, especially as common names are used on some syllabuses.

Quite recently, two new and very important terms have come into general use. These are Prokaryota (prokaryotes, procaryotes) and Eukaryota (eukaryotes, eucaryotes). The Prokaryota includes those organisms which do not have a true nucleus with a nuclear membrane and whose DNA is not organised into chromosomes. It comprises the bacteria and blue-green algae only. The Eukaryota includes those organisms whose DNA is found in a nucleus, surrounded by a nuclear membrane and organised into chromosomes during cell division. The Eukaryota comprises all organisms except bacteria and blue-green algae, i.e. all other plants and animals. The distinction between pro- and eukaryotes is thought to be the most basic among living organisms. The flagellates, from which plants and animals are thought to have evolved, arose from eukaryotic organisms. Since bacteria and blue-green algae are considered to be closely related, they are often placed together in a division called the Schizophyta, where the blue-green algae are called Schizophyceae and the bacteria are called Schizomycetes.

Besides the great differences between prokaryotes and eukaryotes, some authorities consider that the differences between members of the Eukaryota are so important that there should be more than the two kingdoms of plants and animals. As mentioned above, fungi are often placed in a third and separate kingdom, but a number of classification systems for living organisms have been proposed, with suggestions ranging from three to six kingdoms. One of these suggestions, that of a five-kingdom system, has received much support. This was originally proposed by R. H. Whittaker of Cornell University in papers written in 1959 and 1969, modified by L. Margulis and K. V. Schwartz in 1982, and sometimes called the Margulis 1974 classification.

The five-kingdom system is as follows:

1 Kingdom Monera
Prokaryotic organisms.
(a) Bacteria
(b) Blue-green algae

2 Kingdom Protoctista or Protista
Eukaryotic unicellular organisms and their immediate descendents.
(a) All eukaryotic algae (including seaweeds)
(b) Flagellated fungi, i.e. water moulds, slime moulds
(c) Protozoa

3 Kingdom Fungi
Fungi without flagellated stages.

4 Kingdom Animalia
Multicellular animals including sponges.

5 Kingdom Plantae
Multicellular green plants.
(a) Bryophyta
(b) Pteridophyta
(c) Spermatophyta

There are a number of other classification terms in common use.
Thallophyta: a group whose body is not divided into roots, stems and leaves. Originally this group included algae, fungi, lichens and bacteria, but its constituents have varied in the course of time.
Cormophyta: a group whose body is divided into roots, stems and leaves. This group includes the pteridophytes and spermatophytes, and sometimes the bryophytes.
Archegoniatae: a group whose female sex organs are archegonia. This group always includes bryophytes and pteridophytes, and sometimes the gymnosperms since they have reduced but definite archegonia.
Embryophyta: a group which produce a multicellular embryo within the parent plant. This group includes the bryophytes, pteridophytes and spermatophytes.
Tracheophyta: the plants with vascular tissue (i.e., tracheids). This group includes the pteridophytes and spermatophytes.
Cryptogamia and Phanerogamia: these are two obsolete terms still sometimes seen. Cryptogams are plants which reproduce by spores, and phanerogams are plants which reproduce by seeds.

Division Algae

Characteristics

Photosynthetic plants in which the plant body is unicellular, colonial, thalloid, or parenchymatous, but does not form root, stem and leaves.

Possess a range of carotenoid and biloprotein (phycobilin) accessory pigments which are important in classification, as well as chlorophyll.

Mainly aquatic, marine or fresh water; a few live in damp places on land.

Reproduce by spores.

Sex organs are not surrounded by sterile tissue, i.e. are not archegonia, except in the Charophyceae (which is often included in the Chlorophyceae).

Zygote does not develop into a multicellular embryo inside the female sex organ.

Summary classification of the division

Division Algae
  Class Cyanophyceae
  Class Euglenophyceae
  Class Chlorophyceae
  Class Xanthophyceae
  Class Bacillariophyceae

  Class Chrysophyceae
  Class Pyrrophyceae
  Class Cryptophyceae
  Class Phaeophyceae
  Class Rhodophyceae

Some authorities reduce or increase the number of classes.

In some systems of classification, the Algae is considered to be a group of above division status. The classes are then raised to division level and known as Cyanophyta (Myxophyta), Euglenophyta, Chlorophyta, Xanthophyta, Bacillariophyta, Chrysophyta, Pyrrophyta, Cryptophyta, Phaeophyta, Rhodophyta.

Class Cyanophyceae or Myxophyceae (blue-green algae)

1 Possess chlorophyll and the accessory biloprotein (phycobilin) pigments phycocyanin (blue) and phycoerythrin (red) usually with more phycocyanin, giving the blue colour.
2 Cell wall is made of peptidoglycan (mucopeptide).
3 Food reserves are cyanophycean starch and cyanophycin (protein).
4 Includes unicellular, colonial and filamentous forms.
5 No flagella.
6 Much smaller than other algae.
e.g. *Nostoc, Oscillatoria*.

This group is prokaryotic, while all other algae are eukaryotic. Blue-green algae are thought to be more closely related to bacteria, and may be placed with bacteria in the division Schizophyta. The bacteria are then considered to be a class of Schizophyta called Schizomycetes, and the blue-green algae are a class called Schizophyceae.

Class Euglenophyceae (euglenoids)

1 Possess chlorophyll and carotenoids.
2 Cell wall is absent, and they are also classified as flagellate protozoans.
3 Food reserve is paramylum.
4 Most are unicellular, a few colonial.
e.g. *Euglena*.

Euglenoids are claimed as both plants and animals and are probably intermediate organisms from which plants and animals evolved.
Plant characteristics: (1) photosynthetic; (2) similar to green algae like *Chlamydomonas,* except that they do not have cell walls.
Animal characteristics: (1) can live heterotrophically; (2) similar to flagellates like *Trypanosoma,* except that they can photosynthesise; (3) no cell walls; (4) food reserves not starch.

When classified as animals, euglenoids are placed in the phylum Protozoa, class Flagellata, subclass Phytoflagellata or Phytomastigina.

Class Chlorophyceae (green algae)

1 Possess chlorophyll and some carotenoids, and appear green.
2 Cell wall is usually made of cellulose, but some are without cell walls and are also considered to be protozoans.
3 Food reserve is true starch.
4 Includes unicellular, colonial, filamentous and parenchymatous forms. Some unicellular forms are members of the phytoplankton, especially a group called desmids. Filamentous forms may be rooted, or attached to a surface, or free floating, and larger parenchymatous forms are green seaweeds.
e.g. *Chlorella, Chlamydomonas, Pleurococcus,* which are all unicellular, *Volvox* which is colonial, *Cladophora, Spirogyra,* which are filamentous, *Ulva* (sea lettuce) which is parenchymatous.

The Chlorophyceae is one of the largest groups of algae. It is sometimes divided into two classes, the Chlorophyceae and the Charophyceae. The Charophyceae includes the stoneworts such as *Chara,* which look rather like horsetails and are somewhat different from other algae.

Class Xanthophyceae (yellow-green algae)

1 Possess chlorophyll and many carotenoid pigments, giving them a yellowish colour.
2 Cell wall may be absent, and some members are considered to be flagellate protozoans; if a wall is present, it has much pectic material and is often silicified.
3 Food store is oil or fat.
4 Mostly unicellular or filamentous, some amoeboid; very common in phytoplankton.
e.g. *Botrydium, Chloramoeba.*

Class Bacillariophyceae (diatoms)

1 Possess chlorophyll with several xanthophylls, especially fucoxanthin.
2 Cell wall is made of pectin and silica, and is composed of two halves, overlapping like a date box or petri dish, and called a frustule.
3 Food reserves are chrysolaminarin and fat.
4 Unicellular or colonial; common in phytoplankton.
There are two subgroups:
Order Pennales (pennate diatoms): shaped like a date box, e.g. *Pinnularia.*
Order Centrales (centric diatoms): shaped like a petri dish, e.g. *Coscinodiscus.*

Class Chrysophyceae (golden algae)

1 Possess chlorophyll and the brown pigments fucoxanthin and diadinoxanthin.
2 Cell wall is always absent, but cell membrane may become silicified.
3 Food reserves are oil and chrysolaminarin.
4 Mostly unicellular; very common in phytoplankton.
e.g. *Chromulina* and the coccolithophorids whose skeletons formed the sediment that was compressed to form chalk.

Class Pyrrophyceae (dinoflagellates)

1 Possess chlorophyll and special carotenoids unique to the group.
2 Cell wall is made of plates or valves, or may be absent, and these members are also considered to be protozoans.
3 Food reserves are mannitol and starch.
4 Unicellular, and more rarely palmelloid and filamentous forms; the unicellular forms are called dinoflagellates, and are common in the phytoplankton.
e.g. *Ceratium, Gymnodinium, Peridinium*.

Class Cryptophyceae

This is a very small class which was once included in the Pyrrophyceae. Members usually have no cell walls, and have been considered to be protozoans.
e.g. *Cryptomonas*.

Class Phaeophyceae (brown algae)

1 Possess chlorophyll and the brown pigment fucoxanthin, so appear brown.
2 Cell wall contains alginic acid and fucinic acid.
3 Food reserve is laminarin.
4 Plant body is more complex than in any other group except red algae, and includes filamentous, thalloid and parenchymatous forms; mostly large, marine, brown seaweeds.
e.g. *Ectocarpus*, which is filamentous, *Ascophyllum* (wrack), *Fucus* (wrack), *Laminaria* (kelp or oarweed), *Pelvetia*, all of which are thalloid seaweeds.

Class Rhodophyceae (red algae)

1 Possess chlorophyll and the accessory biloprotein (phycobilin) pigments phycocyanin and phycoerythrin so they appear red.
2 Cell wall includes cellulose and polysulphate esters.
3 Food reserves are floridean starch and floridoside.
4 Usually thalloid, but also unicellular, filamentous and parenchymatous types are known; mainly marine red seaweeds, some impregnated with lime.
5 No flagella (only group of eukaryotic algae with none).
6 Very complex life cycles.
e.g. *Bangia, Chondrus* (carragheen), *Porphyra* (laver), *Rhodymenia* (dulse).

Division	Algae
Class	Cyanophyceae
Order	Nostocales
Genus	Nostoc

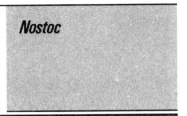

Nostoc

Several trichomes of *Nostoc* embedded in a common mass of mucilage

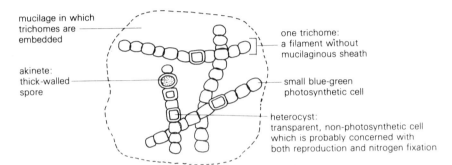

mucilage in which trichomes are embedded

akinete: thick-walled spore

one trichome: a filament without mucilaginous sheath

small blue-green photosynthetic cell

heterocyst: transparent, non-photosynthetic cell which is probably concerned with both reproduction and nitrogen fixation

Simplified diagram of one cell of a trichome of a generalised blue-green alga with mucilage sheath, as seen under electron microscope to show prokaryotic structure

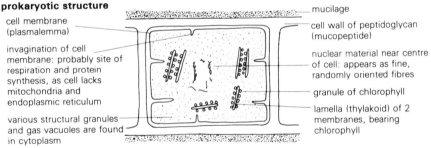

cell membrane (plasmalemma)

invagination of cell membrane: probably site of respiration and protein synthesis, as cell lacks mitochondria and endoplasmic reticulum

various structural granules and gas vacuoles are found in cytoplasm

mucilage

cell wall of peptidoglycan (mucopeptide)

nuclear material near centre of cell: appears as fine, randomly oriented fibres

granule of chlorophyll

lamella (thylakoid) of 2 membranes, bearing chlorophyll

Habitat: ponds, streams, ditches, damp rocks; some species form the algal partner of some lichens, or have a symbiotic association with the roots of certain higher plants, where they are concerned with nitrogen fixation.

Non-sexual reproduction
(1) By fragmentation of filaments; specialised hormogonia (short lengths of trichome with rounded ends) are formed and give rise to new trichomes. In some genera, trichomes are formed by breaks at heterocysts (see glossary).
(2) By special thick-walled spores called akinetes (see glossary).

Sexual reproduction: has not been observed in blue-green algae, but a parasexual process called genetic recombination has been found in some genera.

Notes: *Nostoc* is capable of nitrogen fixation in the soil, converting nitrogen gas to ammonia. Heterocysts may be the site of nitrogen fixation, which probably cannot occur in the presence of oxygen produced in photosynthesis since it requires reducing conditions. Nitrogen fixation by *Nostoc* is important in Asia in rice paddy fields.

In some genera, each trichome is surrounded by mucilage, e.g. *Oscillatoria*; in others there may be many parallel trichomes surrounded by a common sheath, e.g. *Microcoleus*, while in *Nostoc* the trichomes are contorted and embedded in a mucilage mass to form a spherical shape.

Euglena	Division	Algae
	Class	Euglenophyceae
	Order	Euglenales
	Genus	Euglena

L.S. through entire body

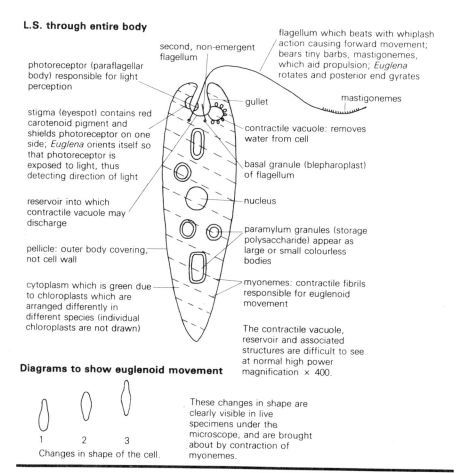

photoreceptor (paraflagellar body) responsible for light perception

second, non-emergent flagellum

flagellum which beats with whiplash action causing forward movement; bears tiny barbs, mastigonemes, which aid propulsion; *Euglena* rotates and posterior end gyrates

mastigonemes

gullet

stigma (eyespot) contains red carotenoid pigment and shields photoreceptor on one side; *Euglena* orients itself so that photoreceptor is exposed to light, thus detecting direction of light

contractile vacuole: removes water from cell

basal granule (blepharoplast) of flagellum

reservoir into which contractile vacuole may discharge

nucleus

paramylum granules (storage polysaccharide) appear as large or small colourless bodies

pellicle: outer body covering, not cell wall

cytoplasm which is green due to chloroplasts which are arranged differently in different species (individual chloroplasts are not drawn)

myonemes: contractile fibrils responsible for euglenoid movement

The contractile vacuole, reservoir and associated structures are difficult to see at normal high power magnification × 400.

Diagrams to show euglenoid movement

1   2   3

Changes in shape of the cell.

These changes in shape are clearly visible in live specimens under the microscope, and are brought about by contraction of myonemes.

Habitat: ponds, ditches and puddles, especially in water containing organic matter.

Reproduction: non-sexual only, by longitudinal binary fission. It can also form a temporary palmella stage (see *Chlamydomonas*, page 15).

Nutrition: in the light, it usually photosynthesises, but in the dark it can undergo saprophytic nutrition, using organic molecules in the water. Unlike most plants, it cannot make vitamins B_1 or B_{12}.

Behaviour: *Euglena* swims towards light of moderate intensity (positive phototaxis) but away from very bright light (negative phototaxis), enabling it to move to optimum light intensity for photosynthesis.

Notes: *Euglena* is considered to be both a plant and an animal, and is probably similar to the ancestral eukaryotic flagellate organisms from which both plants and animals are thought to have evolved (see page 11).

Division	Algae
Class	Chlorophyceae
Order	Volvocales
Genus	*Chlamydomonas*

Chlamydomonas

L.S. through pear-shaped motile unicellular structure

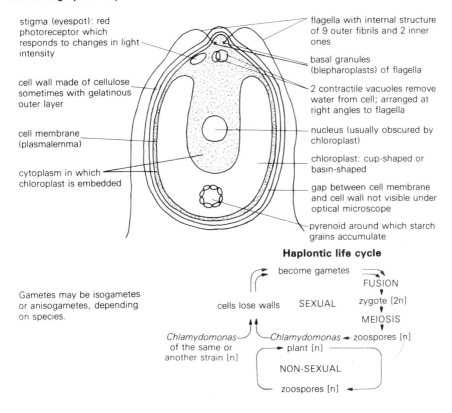

stigma (eyespot): red photoreceptor which responds to changes in light intensity

cell wall made of cellulose sometimes with gelatinous outer layer

cell membrane (plasmalemma)

cytoplasm in which chloroplast is embedded

flagella with internal structure of 9 outer fibrils and 2 inner ones

basal granules (blepharoplasts) of flagella

2 contractile vacuoles remove water from cell; arranged at right angles to flagella

nucleus (usually obscured by chloroplast)

chloroplast: cup-shaped or basin-shaped

gap between cell membrane and cell wall not visible under optical microscope

pyrenoid around which starch grains accumulate

Haplontic life cycle

Gametes may be isogametes or anisogametes, depending on species.

become gametes

FUSION

cells lose walls   SEXUAL   zygote [2n]

MEIOSIS

Chlamydomonas of the same or another strain [n]   *Chlamydomonas* ← zoospores [n]
plant [n]

NON-SEXUAL

zoospores [n]

Habitat: fresh water as part of the phytoplankton. A few species live in the sea and brackish water; one on snow, some on waterlogged soil.

Non-sexual reproduction

(1) By zoospores: flagella are withdrawn and the plant divides into two or more daughter cells within the wall of the parent. Daughter cells develop flagella and cell walls, and are called zoospores. They break free and form new plants.

(2) In unfavourable conditions, particularly when water is scarce, the release of daughter cells is delayed and they go on dividing inside the parent envelope, which becomes mucilaginous. This is called the palmella stage.

Sexual reproduction

Individuals of different (or sometimes the same) mating types come together; they shed their cell walls and become gametes. Then pairs of gametes fuse to form zygotes. Each zygote develops a coat, and undergoes meiosis to form four zoospores which grow into new *Chlamydomonas* plants. This is a haplontic life cycle; only the zygote is diploid, all other stages are haploid.

15

Chlorella	Division	Algae
	Class	Chlorophyceae
	Order	Chlorococcales
	Genus	*Chlorella*

L.S. through non-motile, unicellular, spherical structure

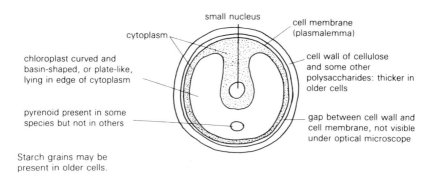

small nucleus

cytoplasm

cell membrane (plasmalemma)

chloroplast curved and basin-shaped, or plate-like, lying in edge of cytoplasm

cell wall of cellulose and some other polysaccharides: thicker in older cells

pyrenoid present in some species but not in others

gap between cell wall and cell membrane, not visible under optical microscope

Starch grains may be present in older cells.

Production of autospores

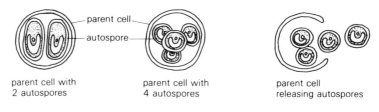

parent cell

autospore

parent cell with 2 autospores

parent cell with 4 autospores

parent cell releasing autospores

Up to 16 autospores may be produced in one parent cell.

Habitat: stagnant water, forming a green suspension; some species live symbiotically in *Hydra* and in some protozoans, e.g. *Stentor*.

Reproduction: is non-sexual only. Cells divide inside the parent wall to form non-motile spores called autospores or aplanospores, which are released by breakdown of the parent wall. Sexual reproduction has not been observed.

Use: (1) *Chlorella* is used in physiological experiments on carbon dioxide in photosynthesis, as it is thought to have chlorophyll similar to that of higher plants. It is good laboratory material in that it is easy to grow, and can be used in easily measurable quantities. (2) Attempts have been made to grow *Chlorella* as food since it has a high protein content, but production is too expensive for commercial exploitation.

Notes: *Chlorella* appears rather similar to *Pleurococcus* in vegetative form, but *Chlorella* is in the order Chlorococcales and *Pleurococcus* is in the Chaetophorales (see *Pleurococcus*, page 17).

Division	Algae	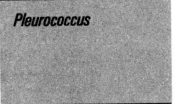
Class	Chlorophyceae	*Pleurococcus*
Order	Chaetophorales	
Genus	*Pleurococcus*	

L.S. through non-motile, unicellular, spherical structure

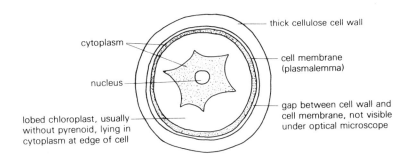

cytoplasm

nucleus

lobed chloroplast, usually without pyrenoid, lying in cytoplasm at edge of cell

thick cellulose cell wall

cell membrane (plasmalemma)

gap between cell wall and cell membrane, not visible under optical microscope

Aggregates of 2 or 4 cells are often seen joined together as a result of binary fission.
In the laboratory, only the cell wall and chloroplast are usually visible at normal high power (× 400) magnification.

Aggregates as seen under high power (× 400) magnification

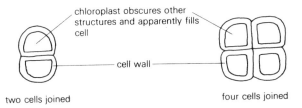

chloroplast obscures other structures and apparently fills cell

cell wall

two cells joined

four cells joined

In very moist conditions, irregular colonies may develop, with short chains of cells joined end to end.

Habitat: growing on tree bark, walls, etc., always in a moist situation; it is one of the few algae to live out of water. Some species form the algal component of certain lichens.
Reproduction: by binary fission only.
Notes: the genus is also called *Protococcus*. Its vegetative form appears similar to that of *Chlorella* (page 16) but the two genera are not closely related. In *Chlorella* the chloroplast is basin-shaped, curved or plate-like while in *Pleurococcus* it is lobed; *Chlorella* lives in fresh water and its reproduction is not by binary fission. *Pleurococcus* is in the order Chaetophorales in which most members are filamentous, and under very moist conditions *Pleurococcus* forms short filaments.

	Division	Algae
Spirogyra blanket weed	*Class*	Chlorophyceae
	Order	Conjugales
	Genus	*Spirogyra*

A few cells of *Spirogyra*

Drawn diagrammatically to show general appearance
of a filament at magnification of × 100.

cell wall — mucilaginous sheath — spiral chloroplast

L.S. through one cell under high magnification

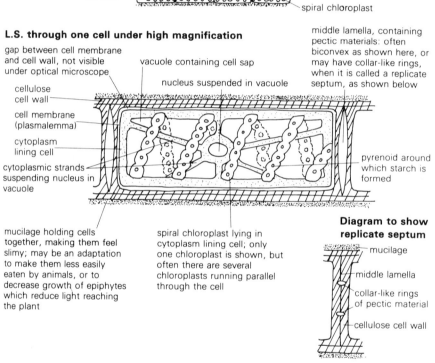

gap between cell membrane and cell wall, not visible under optical microscope

cellulose cell wall

cell membrane (plasmalemma)

cytoplasm lining cell

cytoplasmic strands suspending nucleus in vacuole

vacuole containing cell sap

nucleus suspended in vacuole

middle lamella, containing pectic materials: often biconvex as shown here, or may have collar-like rings, when it is called a replicate septum, as shown below

pyrenoid around which starch is formed

mucilage holding cells together, making them feel slimy; may be an adaptation to make them less easily eaten by animals, or to decrease growth of epiphytes which reduce light reaching the plant

spiral chloroplast lying in cytoplasm lining cell; only one chloroplast is shown, but often there are several chloroplasts running parallel through the cell

Diagram to show replicate septum

mucilage
middle lamella
collar-like rings of pectic material
cellulose cell wall

Habitat: floating in fresh water, sometimes in stagnant water, usually near the surface to obtain light for photosynthesis. A few species are attached to a substratum.

Growth: cells grow to maximum size and then divide transversely by mitosis. All cells can divide. This is intercalary growth.

Non-sexual reproduction: occurs when conditions are good, i.e. warmth and light of summer. It is called **fragmentation.** The filament breaks into short units or separate cells, each of which divides to form a new filament. Fragmentation is caused by the swelling of pectic materials between adjacent cells. In some species, changes in turgor pressure cause the side walls to bulge, pushing the cells apart.

Sexual reproduction: occurs when conditions are bad, i.e. cold and less light of autumn, or absence of nitrates or extreme pH. It is called **conjugation,** see next page.

Notes: *Spirogyra* is recognised by its spiral chloroplast. There are many other genera of filamentous green algae.

Division	Algae
Class	Chlorophyceae
Order	Conjugales
Genus	*Spirogyra*

Spirogyra
blanket weed

sexual reproduction and life cycle

Stages of conjugation in *Spirogyra* (Cells and contents are shown diagrammatically).

(a) Two filaments lie side by side. Produce mucilage which sticks them together.

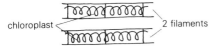

chloroplast — 2 filaments

(b) Protrusions form and join up. Contents round off.

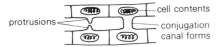

protrusions — cell contents — conjugation canal forms

(c) Contents of one filament move through conjugation canal and fuse with contents of the other filament to form zygotes.

male gamete (moves) [n]

female gamete (does not move) [n]

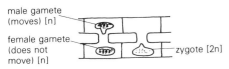

zygote [2n]

(d) This results in one empty filament and one full of zygotes.

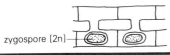

empty filament

zygote [2n] — filament of zygotes

(e) Zygotes become zygospores by development of thick walls.

zygospore [2n]

(f) Filaments sink to bottom of pond and rot, releasing zygospores.

(g) When favourable conditions return, each zygospore germinates into a new filament; it must be immersed in water to do so.

Meiosis occurs when zygospore germinates.

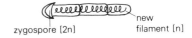

zygospore [2n]   new filament [n]

Haplontic life cycle: anisogamy

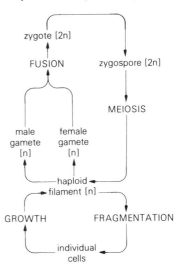

Sexual reproduction is called **conjugation**. It may occur between two filaments (scalariform conjugation) or within one filament (lateral conjugation). Here it is shown between two filaments which come to lie side by side. Protrusions form between cells and join to form conjugation canals. Contents of cells round off from the walls and form gametes. One of the gametes moves through the canal and fuses with the other to form a zygote, which secretes a thick wall and is called a zygospore. The filaments sink to the bottom of the pond and rot, releasing the zygospores. When favourable conditions return, each diploid zygote undergoes meiosis to form four cells, three of which abort while the fourth develops into a new filament. This is a haplontic life cycle; only the zygote is diploid, all other stages are haploid.

Ulva lactuca	Division	Algae
sea lettuce	Class	Chlorophyceae
	Order	Ulvales
	Genus	*Ulva*

Life cycle with isomorphic alternation of generations

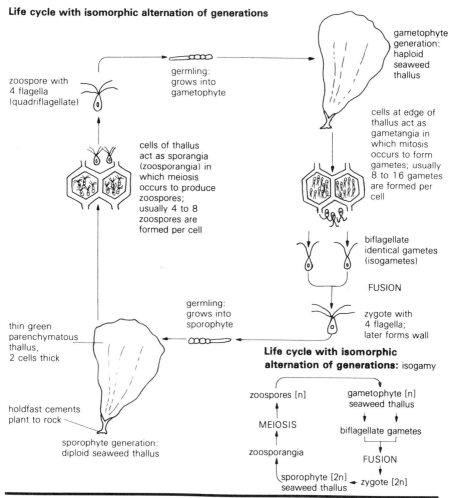

zoospore with
4 flagella
(quadriflagellate)

germling:
grows into
gametophyte

gametophyte
generation:
haploid
seaweed
thallus

cells of thallus
act as sporangia
(zoosporangia) in
which meiosis
occurs to produce
zoospores;
usually 4 to 8
zoospores are
formed per cell

cells at edge of
thallus act as
gametangia in
which mitosis
occurs to form
gametes; usually
8 to 16 gametes
are formed per
cell

biflagellate
identical gametes
(isogametes)

FUSION

germling:
grows into
sporophyte

zygote with
4 flagella;
later forms wall

thin green
parenchymatous
thallus,
2 cells thick

holdfast cements
plant to rock

sporophyte generation:
diploid seaweed thallus

**Life cycle with isomorphic
alternation of generations:** isogamy

zoospores [n] → gametophyte [n]
seaweed thallus

MEIOSIS → biflagellate gametes

zoosporangia

FUSION

sporophyte [2n] ← zygote [2n]
seaweed thallus

Habitat: sea shore of all types, including where fresh water runs into the sea.
Reproduction: *Ulva* has isomorphic alternation of generations, i.e. the sporophyte
and gametophyte look alike. In the sporophyte, any cell of the thallus can act as a
sporangium and produces haploid spores (zoospores) by meiosis. The zoospores
are liberated and swim around for a while before each develops into a several-
celled germling which grows into the haploid gametophyte. Here, any cell can act
as a gametangium and produces isogamous biflagellate gametes. These are
released through a pore in the surface and swim in the water where they fuse to
form a zygote. This swims for a while, then secretes a wall around itself and
remains dormant before growing into a germling and then into the sporophyte.

Division	Algae
Class	Bacillariophyceae
Order	Pennales
Genus	Pinnularia

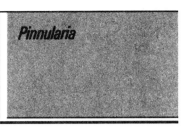

Valve view (view from the top)

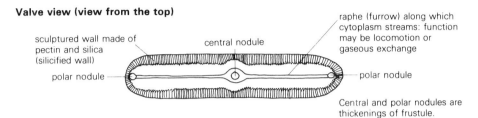

- sculptured wall made of pectin and silica (silicified wall)
- central nodule
- polar nodule
- raphe (furrow) along which cytoplasm streams: function may be locomotion or gaseous exchange
- polar nodule

Central and polar nodules are thickenings of frustule.

Girdle (side) view to show internal structure

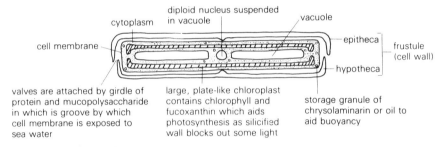

- cytoplasm
- diploid nucleus suspended in vacuole
- vacuole
- cell membrane
- epitheca
- frustule (cell wall)
- hypotheca
- valves are attached by girdle of protein and mucopolysaccharide in which is groove by which cell membrane is exposed to sea water
- large, plate-like chloroplast contains chlorophyll and fucoxanthin which aids photosynthesis as silicified wall blocks out some light
- storage granule of chrysolaminarin or oil to aid buoyancy

Non-sexual reproduction: cells shown diagrammatically

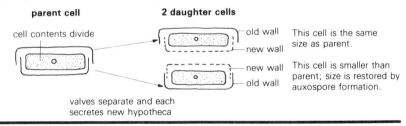

parent cell

2 daughter cells

cell contents divide

- old wall
- new wall

This cell is the same size as parent.

- new wall
- old wall

This cell is smaller than parent; size is restored by auxospore formation.

valves separate and each secretes new hypotheca

Habitat: *Pinnularia* is found in acidic freshwater ponds, ditches and moist soil. Other diatoms are important in marine and freshwater phytoplankton.

Non-sexual reproduction: is by cell division. The two valves of the frustule separate, and each valve forms the epitheca of a daughter cell; new hypothecas are formed within each epitheca, so that one cell is the same size as the mother cell, but the other is smaller. This decrease in size is halted by the production of **auxospores**: the cell contents round off and form wall-less cells called auxospores, which are released. They then enlarge and secrete new walls, restoring the original size.

Sexual reproduction: is by isogamy with diplontic life cycle. Two diploid cells come together and are enclosed in a mucilaginous envelope. Meiosis occurs in each cell to form four haploid cells, three of which abort, and the remaining one acts as a gamete. The gametes are set free from the walls, and fuse to form a zygote, which enlarges to become an auxospore. This develops into a vegetative cell.

	Division	Algae
Fucus species	_Class_	Phaeophyceae
wracks	_Order_	Fucales
	Genus	_Fucus_

Generalised diagram of _Fucus_ to show features of the genus and adaptations to habitat

Dichotomous branching is exaggerated, many branches not shown.

Plant body is a thallus.

growth by apical cell which splits into two, giving dichotomous branching

swollen tip, receptacle, containing fertile conceptacles which contain sex organs

fronds are flattened and contain chlorophyll masked by the brown pigment fucoxanthin which uses light at the blue end of the spectrum (which penetrates to a greater depth) and so aids photosynthesis

midrib of elongated cells for conduction

frond covered with slimy mucilage which:
(i) prevents tearing as fronds slip over one another
(ii) prevents drying out at low tide

fronds are ribbon-like and dissected to prevent tearing by waves

pore (ostiole) leading to sterile conceptacle: little cavity lined with hairs (paraphyses) of 2 types:
(i) do not project from conceptacle and secrete mucilage which oozes out of ostiole to cover frond
(ii) project through ostiole to absorb water and mineral salts from the sea

cylindrical flexible stipe

Ostioles are seen as little dots all over thallus and are not drawn.

hapteron (holdfast) cements plant to rock to prevent dislodging by wave action

hapteron is a different shape in different species

Habitat: intertidal zone of the sea shore, i.e. it is covered at high tide and exposed at low tide. This habitat creates problems: the plant may be ripped by wave action, it may dry out when exposed at low tide, and little light penetrates for photosynthesis. The diagram is annotated with adaptations to habitat.

Notes on the common species of _Fucus_

F. serratus, serrated wrack: has serrated edges to its fronds, no bladders.
F. vesiculosus, bladderwrack: has smooth fronds and air-filled bladders for buoyancy.
F. spiralis, twisted wrack: has smooth, twisted fronds, no bladders.
    The three species grow on different zones of the sea shore; _F. spiralis_ is the highest, then _F. vesiculosus,_ and _F. serratus_ is usually the lowest. There is another species, _F. ceranoides,_ which grows in brackish water, e.g. estuaries.

Reproduction: see page 23.

Division	Algae
Class	Phaeophyceae
Order	Fucales
Genus	Fucus

Fucus species
wracks

life cycle

Life cycle of Fucus

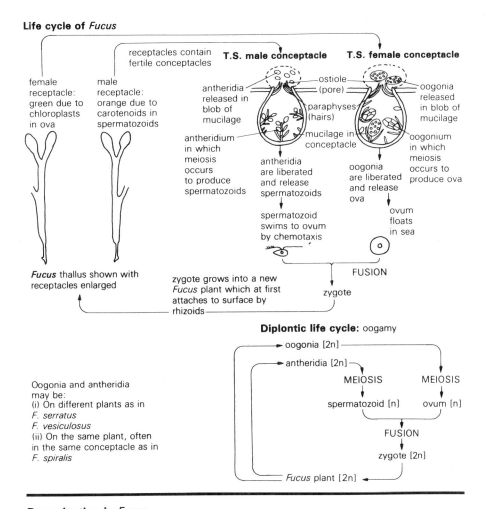

receptacles contain fertile conceptacles

T.S. male conceptacle   **T.S. female conceptacle**

female receptacle: green due to chloroplasts in ova

male receptacle: orange due to carotenoids in spermatozoids

antheridia released in blob of mucilage

ostiole (pore)

paraphyses (hairs)

oogonia released in blob of mucilage

antheridium in which meiosis occurs to produce spermatozoids

mucilage in conceptacle

oogonium in which meiosis occurs to produce ova

antheridia are liberated and release spermatozoids

oogonia are liberated and release ova

spermatozoid swims to ovum by chemotaxis

ovum floats in sea

FUSION

Fucus thallus shown with receptacles enlarged

zygote grows into a new *Fucus* plant which at first attaches to surface by rhizoids

zygote

Diplontic life cycle: oogamy

Oogonia and antheridia may be:
(i) On different plants as in
F. serratus
F. vesiculosus
(ii) On the same plant, often in the same conceptacle as in
F. spiralis

oogonia [2n]

antheridia [2n]

MEIOSIS   MEIOSIS

spermatozoid [n]   ovum [n]

FUSION

zygote [2n]

Fucus plant [2n]

Reproduction in Fucus

Non-sexual reproduction: rare, but a small piece of the plant can break off and settle. This usually happens in still conditions, e.g. salt marshes.

Sexual reproduction: (see diagram) is adapted to the intertidal zone. Swollen tips of the fronds form receptacles containing fertile conceptacles which contain the sex organs, oogonia (female) and antheridia (male) in which meiosis occurs to produce the haploid gametes, ova and spermatozoids. The sex organs are liberated in a blob of mucilage when the fronds dry and shrink at low tide. As the tide comes in, the mucilage and walls of the sex organs dissolve, releasing the gametes. Fusion of gametes occurs in the sea, and the diploid zygote grows into a new *Fucus* plant.

Fucus life cycle is diplontic: the plant is diploid, only the gametes are haploid.

Laminaria species	*Division*	Algae
oarweed, kelp	*Class*	Phaeophyceae
	Order	Laminariales
	Genus	*Laminaria*

Life cycle with heteromorphic alternation of generations

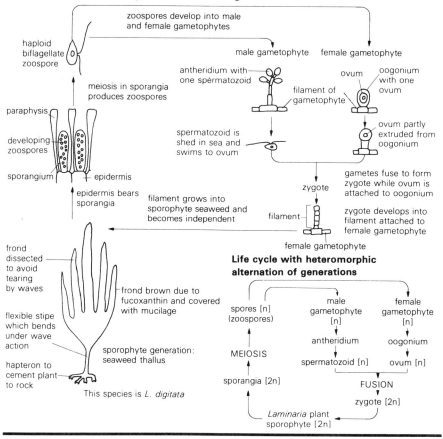

This species is *L. digitata*

Habitat: attached to rocks and stones just below the tide line and exposed only at the spring tides.

Reproduction: *Laminaria* shows heteromorphic alternation of generations; the seaweed is the diploid sporophyte and produces haploid spores by meiosis in sporangia growing from epidermal cells of the thallus. These spores, zoospores, are liberated into the sea by the bursting of the sporangia. They grow into filamentous male and female gametophytes which produce haploid sex organs, antheridia and oogonia, and male and female gametes by mitosis. The male gametes are shed into the sea and swim to the female gametes which are only partially extruded from the oogonia. Here they fuse to form a zygote which grows into the seaweed.

Notes: oarweed is the common name for the genus *Laminaria* and kelp is the name for the order Laminariales. There are three common species of *Laminaria*; *L. saccharina* has strap-like or oar-like fronds, while *L. digitata* and *L. hyperborea* are similar to one another and have fronds divided into ribbons.

Economic importance of algae

Algae can cause both economic gain and economic loss.

Algae causing economic gain

Unicellular and filamentous algae are important members of the phytoplankton, especially diatoms (Bacillariophyceae), desmids (Chlorophyceae) and dinoflagellates (Pyrrophyceae). They are the photosynthetic producers of the sea, forming the basis of aquatic food chains on which the fishing industry ultimately depends. Phytoplankton also produce oxygen as a result of photosynthesis. Since oceans cover two thirds of the world surface, this production is considerable and is dispersed around the world in air currents, contributing to the oxygen supplies of terrestrial life.

There are two algal products commercial importance, agar-agar and alginates.
  Agar-agar is the solid medium used for cultivating bacteria and fungi, which dissolves in hot water and sets as a jelly on cooling. It is produced from various red seaweeds, including species of *Gelidium, Gigartina* and *Gracilaria.* Agar is also used in fish canning, as a thickening agent for ice cream, as a laxative, in the manufacture of glue and paper, in the finishing of leather goods and in the sizing of fabrics, and in wine-making and brewing as a clearing agent.
  Alginates are the salts of alginic acid and are produced from brown seaweeds, especially kelps (order Laminariales). They are gelatinous substances with many and varied uses, including soil improvers, stabilising ice cream, thickening soups and sauces, making artificial silks and plastics, and making transparent wrappings of various kinds, such as sausage skins. Alginic acid can prevent bleeding, and gauze containing calcium alginate is used as a surgical dressing.

Some algae are eaten. Their nutritional content is in doubt because, although they contain protein, it is rather indigestible to people.
  The purple laver, *Porphyra umbilicalis,* is used in Britian to make laver bread; the plants are boiled to produce a jelly which is covered with oatmeal and fried. In Ireland it is called ''sloke'' and is sometimes seen in fishmongers. In Japan, where seaweeds are more widely eaten, *Porphyra tenera* is cultivated on seaweed farms, and is eaten in soups, as a covering for balls of rice, and as a pickle. A food called *kombu* is made from several genera of kelps, and other seaweeds are also eaten in China and Japan, but their main use is probably to add roughage to a diet consisting mainly of rice and fish. In Britain, seaweeds eaten also include Irish moss or carragheen, *Chondrus crispus,* which is used as a food, and an extract called carrageenin is used as a medicine against coughs and colds, and as an emulsifier in making jellies, blancmanges, salad creams, etc. The U.S.A. is one of the main commercial sources, but carragheen is also collected in Brittany and Ireland. Dulse, *Rhodymenia palmata,* is sometimes eaten as a salad, or cooked, or used as a chewing gum. In New England it is eaten as a pickle. The blue-green alga, *Nostoc,* is eaten in China.

Seaweed can also be used as animal fodder, and herbivores seem better able to digest it than man. On the Orkney and Shetland islands, a race of sheep which are important in the knitwear industry there, feed largely or on some islands entirely on seaweeds. Since 1979 the Shetland sheep have been threatened by oil pollution which has contaminated their food supply.

The fossilised shells of certain algae have their uses.
  Diatom shells sink to the bottom of the ocean or lake, forming a deposit called

25

diatomaceous earth, diatomin, or kieselguhr. It is used in a variety of ways: traditionally it makes a fine abrasive found in toothpastes and silver polishes, and was formerly used to absorb liquid nitro-glycerine in making dynamite. Today it is used mainly as an insulator and for filtering liquids, such as in refining sugar.

Chalk is also formed from fossilised algae; it is made of shells, called coccoliths, of a group of golden algae, the coccolithophorids.

Some blue-green algae such as *Nostoc* can fix nitrogen in the soil, making crops independent of nitrate fertilisers. *Nostoc* is particularly important in nitrogen fixation for rice, and is a very common micro-organism in paddy fields. Work is being done in the U.S.A. to develop strains of *Nostoc* that will work on other crops in ordinary fields instead of watery paddies.

In the past, brown algae, especially kelps, have been burnt to obtain various products, particularly soda, potash and iodine, but today other sources of these substances are more readily available. Seaweeds have also been used as manure, since they contain nitrogen, potash, phosphorus and trace elements, but they are lower in nitrogen and phosphorus than farmyard manure.

Algae causing economic loss

A few types of planktonic algae are harmful when large quantities of phytoplankton known as "algal blooms" are produced; some of these can be poisonous because of toxins made by the algae.

One of the most spectacular algal blooms is known as the Red Tide, and is produced from time to time off the coast of Florida. The water is turned red by a dinoflagellate, *Gymnodinium brevis,* which contains a red pigment. The dinoflagellate gives off a toxin that poisons fish swimming in it; porpoises, turtles, and birds eating the fish are also poisoned, as are clams and oysters which feed on the plankton. A gas is given off and carried inland causing people to cough and sneeze.

Some planktonic algae can be poisonous if ingested with drinking water or during swimming, and can cause gastric or skin infections and respiratory problems.

Algal blooms also occur in fresh water, especially when there are large quantities of minerals in the water. This can happen if fertilisers applied to the soil in excessive quantities make their way into rivers and streams. It occurs when treated sewage which is rich in phosphates is returned to rivers. In drought conditions as rivers begin to dry up, minerals such as nitrates and phosphates become concentrated in them, encouraging algal blooms. The blooms are harmful because the algae die and are digested by bacteria which take oxygen from the water, so that less is available to fish.

Algae can be troublesome during water purification. Water reaching homes must be pure and free from micro-organisms, so algae are filtered out in the filter beds. Some members of the Xanthophyceae and Cyanophyceae give an unpleasant taste to the water. Here filtration is ineffective and chlorination or absorption with activated charcoal is necessary. When large quantities of algae are present, filters may become clogged. To reduce algae problems, algicides such as copper sulphate or potassium permanganate are used in reservoirs.

Division Fungi

Characteristics

Non-photosynthetic plants with the body usually made of filaments called hyphae, forming a cotton-wool-like mass, the mycelium. In higher fungi the hyphae are massed together in fruit bodies (also called sporocarps or sporophores).

Hyphae usually have cell walls of chitin, although sometimes of cellulose, and may be septate, i.e. with cross walls, or non-septate (coenocytic).

All are non-photosynthetic plants without chlorophyll, so their nutrition is heterotrophic, either saprophytic, parasite, or symbiotic.

Reproduction is by spores.

Summary classification of the division

Division Fungi
  Class Phycomycetes
  Class Ascomycetes
  Class Basidiomycetes
  Class Deuteromycetes (Fungi Imperfecti)

There is another group of fungi, the slime moulds, known as **Myxomycetes,** which are also classified as protozoan animals and called Mycetozoa. They do not have cell walls, except in the spores, and the body is of amoeba-like form, or a mass of amoebae clumped together forming a structure called a plasmodium. They are found in damp places on land.

Class Phycomycetes

1 Hyphae non-septate, and walls are occasionally of cellulose.
2 Non-sexual reproduction by aplanospores.
3 Sexual reproduction usually by oospores (in Oomycetes) or zygospores (in Zygomycetes).
4 Do not form fruit bodies.
e.g. *Mucor, Rhizopus* (pin or bread moulds), *Peronospora* (downy mildew), *Phytophthora* (potato blight).
   The Phycomycetes are also called the lower fungi or algal fungi. In some classification systems the class is split up into six classes, the Chytridiomycetes, Hyphochytridiomycetes, Oomycetes, Zygomycetes, Trichomycetes and Plasmodiophoromycetes.

Class Ascomycetes

1 Hyphae septate.
2 Reproduction is by ascospores borne *inside* a structure called an ascus.
3 Asci may be organised into fruit bodies called ascocarps.

Subgroups include:

(a) Hemiascomycetes
   Asci are not grouped into ascocarps.
   e.g. *Saccharomyces* (yeast).

(b) Plectomycetes
   Ascocarp is a closed, spherical structure called a cleistothecium.

e.g. *Aspergillus, Erysiphe* (powdery mildew). *Penicillium* may be placed here.

(c) Pyrenomycetes (flask fungi)
Ascocarp is a flask-shaped structure called a perithecium.
e.g. *Claviceps* (ergot), *Sordaria, Ceratocystis* (Dutch elm disease) which may be placed in the Plectomycetes.

(d) Discomycetes (cup fungi)
Ascocarp is a cup- or disc-shaped body called an apothecium.
e.g. *Peziza* (elf cap), *Morchella* (morel), *Tuber* (truffle).

Class Basidiomycetes

1 Hyphae septate, often with dolipore septa.
2 Reproduction is by basidiospores borne on the *outside* of a structure called a basidium.
3 Basidia may be organised into complex fruit bodies called basidiocarps.

Subgroups include:

(a) Teliomycetes
No basidiocarp
   Order Uredinales: rusts, e.g. *Puccinia.*
   Order Ustilaginales: smuts, e.g. *Ustilago.*

(b) Hymenomycetes
Basidiocarp present, hymenium exposed at maturity.
   Order Tremellales: jelly fungi, e.g. *Auricularia* (Jew's ear).
   Order Agaricales: agarics (gill fungi), mushrooms, toadstools,
   e.g. *Agaricus.*
   Order Polyporales: polypores (pore fungi), e.g. *Polyporus, Boletus.*
In some classifications, the Polyporales are included in the Agaricales. In other systems, fungi with a leathery fruit body, usually bracket fungi, are placed in the order Aphyllophorales, while those with a fleshy fruit body, whether with gills or pores, are placed in the Agaricales.

(c) Gastromycetes
Basidiocarp present, hymenium not exposed at maturity.
   Order Lycoperdales: puffballs, e.g. *Lycoperdon.*
   Order Sclerodermales: earth stars, e.g. *Scleroderma.*
   Order Nidulariales: bird's nest fungi, e.g. *Nidularia.*
   Order Phallales: stinkhorns, e.g. *Phallus.*

Class Deuteromycetes (Fungi Imperfecti)

1 Hyphae usually septate.
2 No sexual reproduction, so cannot be attributed to any other group. They are probably descended from Ascomycetes or Basidiomycetes which have lost their powers of sexual reproduction.
   e.g. *Penicillium,* which is probably descended from an ascomycete.

Note: subgroups are given the status of subclasses on the annotated diagrams, but the official subclass ending of -mycetidae is not used, since the level of these subgroups can vary in different classification systems.

Division	Fungi

Nutrition and general structure

Comparison of saprophytes and parasites

Saprophyte, e.g. *Rhizopus, Penicillium*	Parasite, e.g. *Phytophthora*
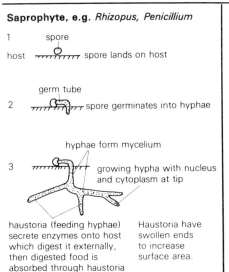	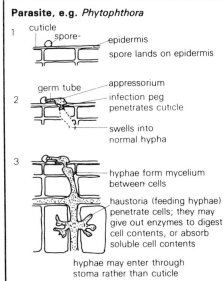

Saprophyte side:

1 spore
host — spore lands on host

2 germ tube — spore germinates into hyphae

3 hyphae form mycelium
growing hypha with nucleus and cytoplasm at tip

haustoria (feeding hyphae) secrete enzymes onto host which digest it externally, then digested food is absorbed through haustoria into mycelium

Haustoria have swollen ends to increase surface area.

Parasite side:

1 cuticle
spore-
epidermis
spore lands on epidermis

2 germ tube — appressorium
infection peg penetrates cuticle
swells into normal hypha

3 hyphae form mycelium between cells

haustoria (feeding hyphae) penetrate cells; they may give out enzymes to digest cell contents, or absorb soluble cell contents

hyphae may enter through stoma rather than cuticle

Generalised structure of hypha, L.S.

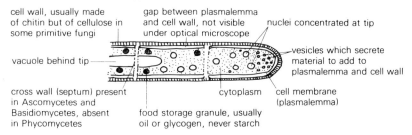

cell wall, usually made of chitin but of cellulose in some primitive fungi

gap between plasmalemma and cell wall, not visible under optical microscope

nuclei concentrated at tip

vacuole behind tip

vesicles which secrete material to add to plasmalemma and cell wall

cross wall (septum) present in Ascomycetes and Basidiomycetes, absent in Phycomycetes

cytoplasm

cell membrane (plasmalemma)

food storage granule, usually oil or glycogen, never starch

Saprophytic fungi contain enzymes which can digest soluble long-lasting polysaccharides, proteins, etc. But saprophytes do not possess a mechanism for entering the host or resisting the host defences.

Parasitic fungi do not have enzymes capable of digesting stable compounds and require their food as unstable monosaccharides or sometimes even organic acids such as those of the Krebs' cycle, which break down when the plant dies. But parasites have a method of entering the host, often a germ tube and appressorium. The mycelium usually lives between cells (intercellular) but haustoria penetrate the cells (intracellular) to absorb food. Some parasites can live as saprophytes if the host dies (facultative parasites), but some require such unstable intermediates that they die if the host dies (obligate parasites).

29

Rhizopus stolonifer	Division	Fungi
black pin mould	Class	Phycomycetes
	Subclass	Zygomycetes
	Order	Mucorales
	Genus	Rhizopus

General structure of *Rhizopus*

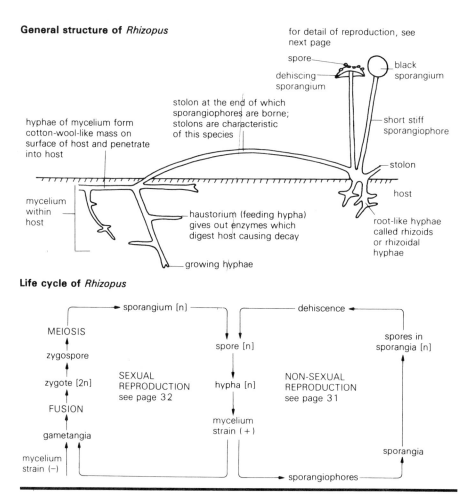

for detail of reproduction, see next page

spore

dehiscing sporangium

black sporangium

stolon at the end of which sporangiophores are borne; stolons are characteristic of this species

hyphae of mycelium form cotton-wool-like mass on surface of host and penetrate into host

short stiff sporangiophore

stolon

host

mycelium within host

haustorium (feeding hypha) gives out enzymes which digest host causing decay

root-like hyphae called rhizoids or rhizoidal hyphae

growing hyphae

Life cycle of *Rhizopus*

sporangium [n]

dehiscence

MEIOSIS

zygospore

spore [n]

spores in sporangia [n]

zygote [2n]

SEXUAL REPRODUCTION see page 32

hypha [n]

NON-SEXUAL REPRODUCTION see page 31

FUSION

gametangia

mycelium strain (+)

sporangia

mycelium strain (−)

sporangiophores

Habitat: *Rhizopus stolonifer* is the black pin mould fungus that grows on bread, and was formerly called *Mucor stolonifer*. It is the genus often called "*Mucor*" in elementary biology courses.

Notes: *Rhizopus* and *Mucor* and very similar, but *Rhizopus* produces its sporangiophores at the ends of special long hyphae called stolons and above clumps of root-like hyphae called rhizoids. The method of dehiscence of the sporangia is also different (see page 31). *Mucor* grows best in horse dung and soil, but species of *Rhizopus* and *Mucor* also grow saprophytically on stale food. *Mucor* will grow on bread, but less readily than *Rhizopus*. *Rhizopus* grows on apples and other fruits in storage and can be grown easily on apple cores in the laboratory. Bread is often an unsuccessful host because mass-produced bread includes a preservative which inhibits pin mould growth.

Division	Fungi
Class	Phycomycetes
Subclass	Zygomycetes
Order	Mucorales
Genus	Rhizopus

Rhizopus stolonifer
black pin mould

non-sexual reproduction

Stages of non-sexual reproduction: sporangia shown in longitudinal section

1 Immature sporangia

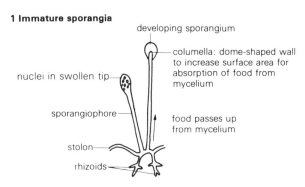

developing sporangium

columella: dome-shaped wall to increase surface area for absorption of food from mycelium

nuclei in swollen tip

sporangiophore

food passes up from mycelium

stolon

rhizoids

Vertical hyphae (sporangiophores) grow upwards from the stolon. The ends swell, and many nuclei migrate into the swelling. Groups of nuclei become surrounded by cytoplasm, forming a spore, and the swollen hypha develops into a sporangium.

2 Mature sporangium

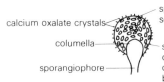

spore containing several nuclei

calcium oxalate crystals

columella

sporangiophore

sporangium appears black due to crystals of calcium oxalate in the wall, produced by metabolism of fungus

The sporangium matures and spores separate. Non-motile spores like this are called aplanospores.

3 Dehiscing sporangium

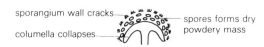

sporangium wall cracks

columella collapses

spores forms dry powdery mass

The sporangium wall dries and cracks into pieces, and columella collapses. Spores and wall fragments are spread over surface. Spores form a dry powdery mass and are dispersed in air currents.

If a spore lands on a suitable material, it germinates to form a germ tube which penetrates host and starts to secrete enzymes, absorb food, and grow.

4 Spore germination

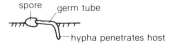

spore   germ tube

hypha penetrates host

Non-sexual reproduction is often the only kind that occurs in the wild. Spores are produced by mitosis, so give fungi identical to the parent. Reproduction in *Mucor* is similar, except that the sporangium wall liquifies at maturity and the spores become stuck to the columella in dry air; in wet air dispersal of spores occurs if a rain drop hits or an insect walks over the sporangium. *Rhizopus* spores are common in air, but those of *Mucor* are rarer.

Rhizopus stolonifer	Division	Fungi
black pin mould	Class	Phycomycetes
	Subclass	Zygomycetes
	Order	Mucorales
sexual reproduction	Genus	*Rhizopus*

Stages of conjugation

1   (+)   (−)

Two hyphae of different strains (+) and (−) come together.

2   progametangia

The tips of hyphae swell up to form progametangia and the cytoplasm and nuclei migrate into the tip.

3   gametangia

suspensor

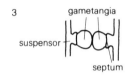

septum

Each tip is cut off by a cross wall (septum) and is now called a gametangium.

4   zygote

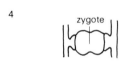

The dividing wall between gametangia breaks down; the contents, including nuclei, fuse to form a zygote, so that the nuclei are diploid and contain (+ −). It is not clear whether all pairs of nuclei fuse in the gametangia or whether only one pair does so.

5   zygospore

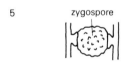

The zygote becomes surrounded by a thick wall and is called a zygospore. It can withstand adverse conditions.

6

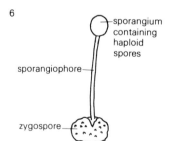

sporangium containing haploid spores

sporangiophore

zygospore

When good conditions return, the zygospore germinates to give a sporangium and spores. This sporangium is called a germ sporangium.

The spores are haploid but it is not clear where meiosis occurs. Since the diploid zygote was (+ −) we would expect 50% of the spores to be (+) and 50% to be (−). But all spores are either (+) or (−) so one type is thought to abort after meiosis.

Sexual reproduction is called **conjugation** and is rare in the wild, although it tends to happen when conditions for growth become poor. It may occur between mycelia of two different strains (heterothallism), or within the same mycelium or two identical mycelia (homothallism). Heterothallism is more common, but homothallism is found in one species, *Rhizopus sexualis*.

Division	Fungi
Class	Phycomycetes
Subclass	Oomycetes
Order	Peronosporales
Genus	Phytophthora

Phytophthora infestans
potato blight

L.S. through host with fungus

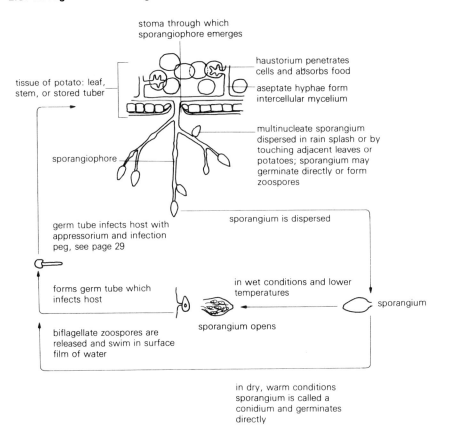

stoma through which sporangiophore emerges

haustorium penetrates cells and absorbs food

aseptate hyphae form intercellular mycelium

tissue of potato: leaf, stem, or stored tuber

multinucleate sporangium dispersed in rain splash or by touching adjacent leaves or potatoes; sporangium may germinate directly or form zoospores

sporangiophore

germ tube infects host with appressorium and infection peg, see page 29

sporangium is dispersed

in wet conditions and lower temperatures

forms germ tube which infects host

sporangium opens

sporangium

biflagellate zoospores are released and swim in surface film of water

in dry, warm conditions sporangium is called a conidium and germinates directly

---

Habitat and nutrition: parasitic on all parts of potato, including the growing plant and stored tubers.

Reproduction: is mainly non-sexual. The sexual stage is rare and requires two different strains (heterothallism). In North America, Europe, and South Africa, only one strain is present, so reproduction is non-sexual only. Where sexual reproduction occurs, oospores are produced.

Economic importance: *Phytophthora infestans* causes the disease potato blight which weakens the growing plant so that tubers are not produced, and rots tubers in storage, making them inedible. It is an endemic disease in Britain, but reached epidemic proportions in Ireland in the 1840s, causing the Irish potato famine. This was at its worst in 1845, when it is estimated that half of the population of Ireland died or emigrated to mainland Britain or the U.S.A.

Saccharomyces species	Division	Fungi
yeast	Class	Ascomycetes
	Subclass	Hemiascomycetes
	Order	Endomycetales
	Genus	Saccharomyces

Structure shown in longitudinal section

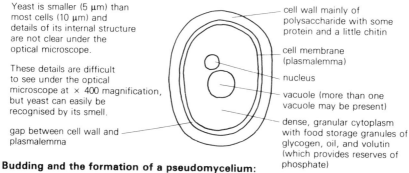

Yeast is smaller (5 µm) than most cells (10 µm) and details of its internal structure are not clear under the optical microscope.

These details are difficult to see under the optical microscope at × 400 magnification, but yeast can easily be recognised by its smell.

gap between cell wall and plasmalemma

cell wall mainly of polysaccharide with some protein and a little chitin

cell membrane (plasmalemma)

nucleus

vacuole (more than one vacuole may be present)

dense, granular cytoplasm with food storage granules of glycogen, oil, and volutin (which provides reserves of phosphate)

Budding and the formation of a pseudomycelium:
only outline of cell and nucleus shown

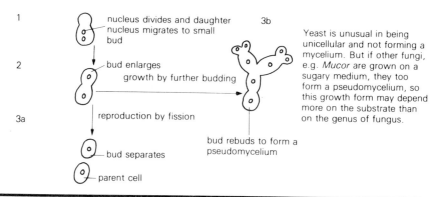

1

nucleus divides and daughter nucleus migrates to small bud

2

bud enlarges

growth by further budding

3a

reproduction by fission

bud separates

parent cell

3b

bud rebuds to form a pseudomycelium

Yeast is unusual in being unicellular and not forming a mycelium. But if other fungi, e.g. *Mucor* are grown on a sugary medium, they too form a pseudomycelium, so this growth form may depend more on the substrate than on the genus of fungus.

Habitat: wild yeast is found on decaying fruits. Cultivated yeasts are *Saccharomyces cerevisiae* (baker's and brewer's yeast) and *S. ellipsoideus* (sometimes considered a strain of *S. cerevisiae*) which is used in wine-making, either as a wild yeast forming a bloom on grape skin or added to the grape juice.
Nutrition: yeast is considered to be a saprophyte, but it cannot use polysaccharide as a food source. It does not possess the enzyme amylase and so cannot survive on starch. In baking, sugar (sucrose) is added to the yeast, and there is some amylase in flour. In brewing, barley is first sprouted to make its own amylase soluble, which then converts starch to maltose. Yeast possesses invertase and maltase, so it can use sucrose and maltose to make hexose sugar for respiration.
Economic importance: The respiration of yeast is used in making bread and alcohol. In bread-making, bubbles of carbon dioxide are trapped in dough to make bread rise. Here the respiration may be aerobic or anaerobic. In alcohol-making, yeast must respire anaerobically and alcohol and carbon dioxide are produced as waste products of this anaerobic respiration (fermentation). Yeast is also used in the production of vitamin supplements (yeast-vite, marmite), as it is rich in vitamins of the B complex. It is a source of single-cell protein.

Division	Fungi
Class	Ascomycetes
Subclass	Hemiascomycetes
Order	Endomycetales
Genus	Saccharomyces

Saccharomyces species
yeast

reproduction

Sexual reproduction and life cycle in yeast:
outline of cell and nucleus only are shown

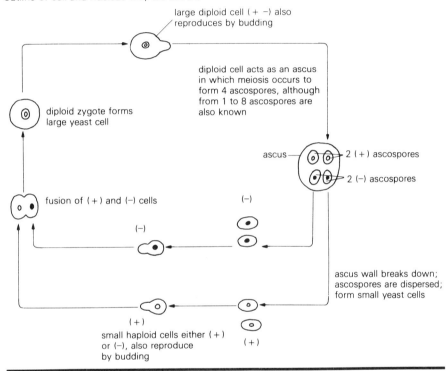

large diploid cell (+ −) also reproduces by budding

diploid cell acts as an ascus in which meiosis occurs to form 4 ascospores, although from 1 to 8 ascospores are also known

diploid zygote forms large yeast cell

ascus — 2 (+) ascospores

2 (−) ascospores

fusion of (+) and (−) cells

(−)

(−)

ascus wall breaks down; ascospores are dispersed; form small yeast cells

(+)

small haploid cells either (+) or (−), also reproduce by budding

(+)

Non-sexual reproduction: is often said to be by budding. But budding is actually just the production of a protuberance known as a bud (as shown on page 34); the nucleus divides first, and one daughter nucleus migrates to an outpushing of the cytoplasm, forming a bud. At this stage one of two things may happen:
(1) The bud may separate from the parent cell forming a new plant; this is non-sexual reproduction by fission, and is sometimes called vegetative reproduction or reproduction by budding.
(2) More buds may be made, forming a pseudomycelium. This is growth, not reproduction.
Budding only occurs in good conditions, i.e. plenty of food, moisture, oxygen, and the absence of toxic conditions.
Sexual reproduction: is by the production of ascospores. The diploid (+ −) cell undergoes meiosis, and the four products of meiosis are ascospores, with the cell acting as an ascus (sporangium). Ascospores are liberated and grow into haploid cells. Yeast is heterothallic: two haploid cells of different strains (+) and (−) fuse to form the diploid cell (+ −). Since there are both haploid and diploid forms, each giving rise to the other, yeast shows alternation of generations which is unusual in fungi. Sexual reproduction occurs in poor conditions, i.e. dry, little food or oxygen.

Aspergillus	Division	Fungi
	Class	Ascomycetes
	Subclass	Plectomycetes
	Order	Eurotiales
	Genus	*Aspergillus*

Part of *Aspergillus* **with conidial head in L.S.**

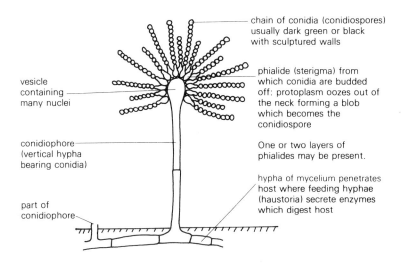

chain of conidia (conidiospores) usually dark green or black with sculptured walls

phialide (sterigma) from which conidia are budded off: protoplasm oozes out of the neck forming a blob which becomes the conidiospore

One or two layers of phialides may be present.

hypha of mycelium penetrates host where feeding hyphae (haustoria) secrete enzymes which digest host

vesicle containing many nuclei

conidiophore (vertical hypha bearing conidia)

part of conidiophore

Entire conidial head, to reduced scale

conidia form a spherical head which can resemble *Rhizopus* and *Mucor* to the naked eye

Section through cleistothecium

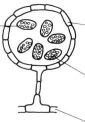

ascus containing 8 ascospores: asci break and ascospores lie free in cleistothecium and are dispersed when cleistothecium decays

cleistothecium made of massed hyphae, often yellow in colour

ascospore which germinates into a new mycelium after dispersal

hypha of mycelium

Habitat and nutrition: similar to *Penicillium* (see page 37).

Reproduction: is both non-sexual with the production of conidiospores, and sexual with the production of ascospores in asci formed in a cleistothecium. The sexual stage was formerly called *Eurotium.*

Economic importance: *Aspergillus* causes decay in stored products by its saprophytic nutrition, but some species are used in the biosynthesis of various organic substances, e.g. citric acid, gluconic acid, gallic acid, cortisone. One species lives in mouldy hay and causes the disease "farmers' lung", apergillosis.

Notes: *Aspergillus* has black spores and can be confused superficially with *Rhizopus* by inexperienced students, since both fungi grow on bread and form a spherical head with black spores, but in *Aspergillus* these are conidiospores borne in chains, while in *Rhizopus* they are aplanospores in sporangia.

36

Division	Fungi
Class	Fungi Imperfecti or Ascomycetes
Order	Hyphomycetales*
Genus	Penicillium

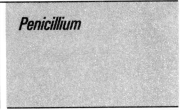

Generalised diagram to show features of the genus *Penicillium*

Species vary in size, length of chains of conidia and size of penicillus.

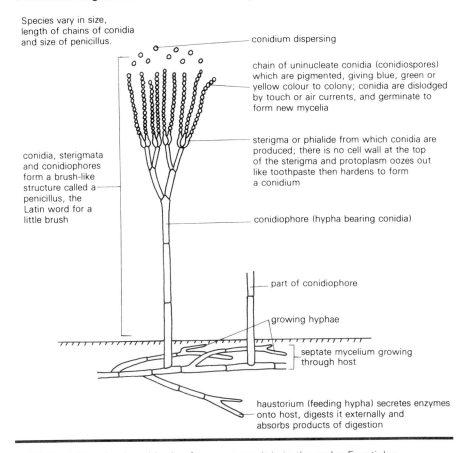

conidium dispersing

chain of uninucleate conidia (conidiospores) which are pigmented, giving blue, green or yellow colour to colony; conidia are dislodged by touch or air currents, and germinate to form new mycelia

conidia, sterigmata and conidiophores form a brush-like structure called a penicillus, the Latin word for a little brush

sterigma or phialide from which conidia are produced; there is no cell wall at the top of the sterigma and protoplasm oozes out like toothpaste then hardens to form a conidium

conidiophore (hypha bearing conidia)

part of conidiophore

growing hyphae

septate mycelium growing through host

haustorium (feeding hypha) secretes enzymes onto host, digests it externally and absorbs products of digestion

*If *Penicillium* is placed in the Ascomycetes it is in the order Eurotiales.
Habitat: bread, cheese, fruit, soil, decaying vegetable matter.
Nutrition: saprophytic and capable of breaking down a range of stable compounds including starches, proteins and fats, but not cellulose so it does not grow on wood.
Notes: most species of *Penicillium* have no sexual reproduction so are placed in the Fungi Imperfecti, but similar genera such as *Aspergillus* also reproduce by ascospores, so *Penicillium* is sometimes placed in the Ascomycetes.
Economic importance: *Pencillium notatum* is the species in which the antibiotic penicillin was first discovered by Sir Alexander Fleming in 1928–9. *P. roqueforti* and *P. camemberti* are important in ripening some cheeses. Many species cause spoilage of textiles and stored food. The green moulds on bread, jam, oranges, etc. are species of *Penicillium*.

Division	Fungi
Class	Ascomycetes
Subclass	Discomycetes
Order	Pezizales
Genus	*Peziza*

Peziza species

elf cap

Structure of *Peziza*

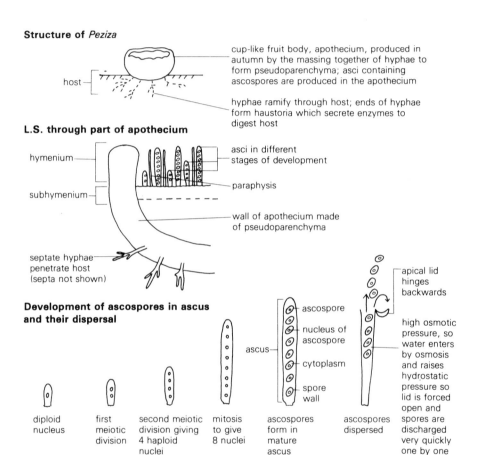

host

cup-like fruit body, apothecium, produced in autumn by the massing together of hyphae to form pseudoparenchyma; asci containing ascospores are produced in the apothecium

hyphae ramify through host; ends of hyphae form haustoria which secrete enzymes to digest host

L.S. through part of apothecium

hymenium

subhymenium

asci in different stages of development

paraphysis

wall of apothecium made of pseudoparenchyma

septate hyphae penetrate host (septa not shown)

Development of ascospores in ascus and their dispersal

ascospore

nucleus of ascospore

ascus

cytoplasm

spore wall

apical lid hinges backwards

high osmotic pressure, so water enters by osmosis and raises hydrostatic pressure so lid is forced open and spores are discharged very quickly one by one

| diploid nucleus | first meiotic division | second meiotic division giving 4 haploid nuclei | mitosis to give 8 nuclei | ascospores form in mature ascus | ascospores dispersed | |

Habitat: many species of *Peziza* are found on dead trees and in soil rich in humus. *Peziza* (or *Aleuria*) *aurantia*, the orange peel peziza, can grow from 1 to 12 cm in diameter and is found on bare gravel, paths, bare soil in woods, and on lawns.

Nutrition: saprophytic; haustoria secrete enzymes including those which can digest wood, e.g. cellulases, causing wood to decay.

Reproduction: fruit bodies are ascocarps called apothecia which produce asci (sporangia containing ascospores) on a layer called the hymenium. When young, the ascus contains a single diploid nucleus which has been formed by a very complex process of sexual reproduction. The nucleus undergoes meiosis to form a line of four nuclei, which then divide by mitosis to form a line of eight nuclei. Each becomes surrounded by cytoplasm and a wall, and is an ascospore. The mature asci develop high osmotic pressure; when in contact with moist air, water is taken in and the ascus bursts by breaking of the lid, which hinges backwards. Spores are dispersed in a cloud. Each germinates to form hyphae in the wood.

Division	Fungi
Class	Basidiomycetes
Subclass	Hymenomycetes
Order	Agaricales
Genus	*Agaricus*

Agaricus bisporus
cultivated mushroom

External view of stages of cultivated mushroom

Fruit bodies are made of hyphae massed together forming compact tissue called pseudoparenchyma. Fruit bodies start underground and break through compost as they develop.

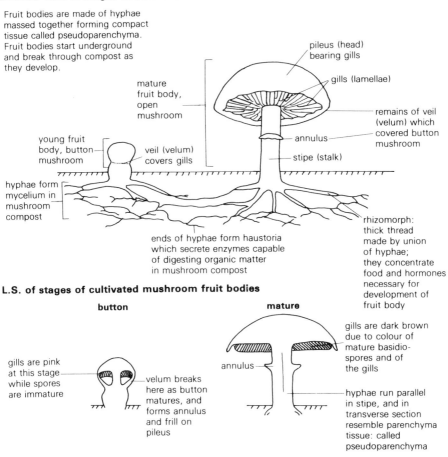

pileus (head) bearing gills

gills (lamellae)

mature fruit body, open mushroom

remains of veil (velum) which covered button mushroom

young fruit body, button mushroom

veil (velum) covers gills

annulus

stipe (stalk)

hyphae form mycelium in mushroom compost

rhizomorph: thick thread made by union of hyphae; they concentrate food and hormones necessary for development of fruit body

ends of hyphae form haustoria which secrete enzymes capable of digesting organic matter in mushroom compost

L.S. of stages of cultivated mushroom fruit bodies

button

mature

gills are pink at this stage while spores are immature

velum breaks here as button matures, and forms annulus and frill on pileus

annulus

gills are dark brown due to colour of mature basidiospores and of the gills

hyphae run parallel in stipe, and in transverse section resemble parenchyma tissue: called pseudoparenchyma

Habitat: the cultivated mushroom, *Agaricus bisporus,* is grown from mycelium "spawn" on mushroom compost containing organic material, often including horse dung, in caves or cellars. When growng wild, it is found on manure heaps and by roadsides. The field mushroom, *Agaricus campestris,* grows in fields rich in dung and occasionally in clearings in woods, but not inside woodland.

Notes: *A. bisporus* is unusual in having only two spores per basidium but most Basidiomycetes, including *A. campestris,* have four.

An "agaric" is any mushroom, toadstool, or bracket fungus with gills (order Agaricales). Some toadstools have pores rather than gills and are called "polypores" (order Polyporales) (see page 42).

The genus *Agaricus* was formerly called *Psalliota.*

Agaricus bisporus cultivated mushroom spore dispersal	Division	Fungi
	Class	Basidiomycetes
	Subclass	Hymenomycetes
	Order	Agaricales
	Genus	Agaricus

Underside of pileus

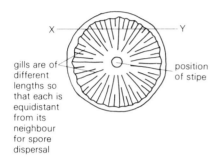

X —— Y

gills are of different lengths so that each is equidistant from its neighbour for spore dispersal

position of stipe

Section across X-Y: not all gills shown

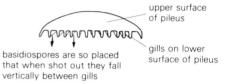

upper surface of pileus

gills on lower surface of pileus

basidiospores are so placed that when shot out they fall vertically between gills

If the pileus is placed on a sheet of paper a spore print is formed. Spores fall between the gills giving a dark brown print, and the gills are left as white areas.

L.S. one gill

hymenium (layer bearing basidia)

subhymenium (layer under hymenium)

trama (inner tissue of gill)

one basidium

region of hymenium with basidia drawn

One basidium

Agaricus bisporus has 2 spores per basidium.

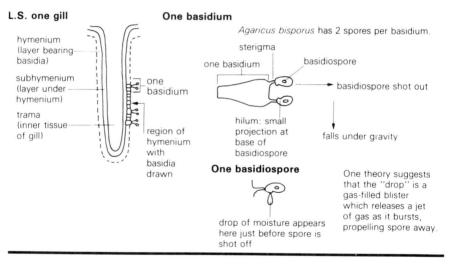

sterigma

one basidium

basidiospore

basidiospore shot out

hilum: small projection at base of basidiospore

falls under gravity

One basidiospore

drop of moisture appears here just before spore is shot off

One theory suggests that the "drop" is a gas-filled blister which releases a jet of gas as it bursts, propelling spore away.

Basidiospores are produced by meiosis in the basidium. The diploid nucleus in the basidium divides to form four haploid nuclei. In most species, each nucleus migrates into one of four fine sterigmata at the end of the basidium and forms a pear-shaped spore, but in *A. bisporus* two abort, so only two spores are formed. The walls of the spores and the gill tissue become coloured with a purplish-brown pigment.

Spores are shot off from the basidia, which act as "spore guns", to a distance equidistant from the two gills, and then fall under gravity. The exact propulsion mechanism is uncertain, but just before dispersal a drop of fluid appears at the junction between the hilum and sterigma, and is carried away with the spore.

For successful spore dispersal, the gills must hang absolutely vertically. This is achieved by negatively geotropic growth of the stipe, and the gills themselves are also sensitive to gravity, and can make small adjustments in their growth so that they are entirely vertical.

Once below the gills, basidiospores are dispersed in air currents.

Division	Fungi
Class	Basidiomycetes
Subclass	Hymenomycetes
Order	Agaricales
Genus	*Agaricus*

Agaricus campestris
field mushroom

life cycle

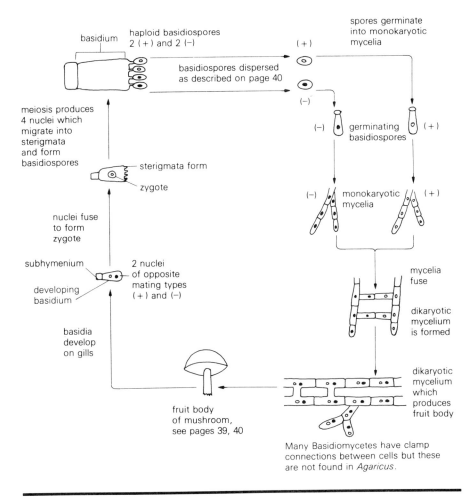

basidium — haploid basidiospores 2 (+) and 2 (−)

basidiospores dispersed as described on page 40

spores germinate into monokaryotic mycelia (+)

(−)

meiosis produces 4 nuclei which migrate into sterigmata and form basidiospores

sterigmata form

zygote

germinating basidiospores (−) (+)

nuclei fuse to form zygote

monokaryotic mycelia (−) (+)

subhymenium — 2 nuclei of opposite mating types (+) and (−)

developing basidium

mycelia fuse

basidia develop on gills

dikaryotic mycelium is formed

fruit body of mushroom, see pages 39, 40

dikaryotic mycelium which produces fruit body

Many Basidiomycetes have clamp connections between cells but these are not found in *Agaricus*.

The mycelium that gives rise to the mushroom fruit body is dikaryotic (i.e. each cell contains two nuclei) which are of different mating types (+) and (−). These two nuclei are also found in the basidia, where they fuse to give a zygote which immediately undergoes meiosis to form four nuclei, two (+) and two (−). In the cultivated mushroom only two nuclei form spores and the other two abort. The nuclei migrate into sterigmata on the basidia and develop into basidiospores. These are dispersed (see page 40) and each grows into a haploid mycelium with one nucleus per cell, called a monokaryotic mycelium. In the soil, different mating types meet, and cells fuse so that each cell contains two nuclei, one of each type, forming a dikaryotic mycelium on which the fruit body develops. Since two different strains are necessary for mating, the mushroom is heterothallic.

Generalised bracket polypore and toadstool polypore		
Division	Fungi
Class	Basidiomycetes
Subclass	Hymenomycetes
Order	Polyporales

Fruit body growing out of tree trunk

One tube

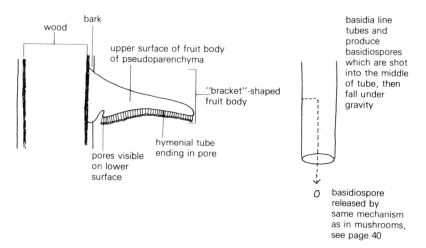

basidia line tubes and produce basidiospores which are shot into the middle of tube, then fall under gravity

O basidiospore released by same mechanism as in mushrooms, see page 40

Fruit body of *Boletus*, a polypore toadstool

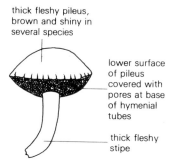

T.S. tube, shown diagrammatically

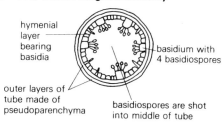

Habitat: bracket fungi may be parasitic on living trees, or saprophytic on dead stumps or branches ; *Boletus* is a mainly woodland genus, whose mycelium forms mycorrhizal associations with roots of various trees, particuarly conifers and birch.
Notes: a fungus with pores rather than gills may be called a polypore (order Polyporales) rather than an agaric (order Agaricales), although some authorities place both groups in the order Agaricales. Most polypores are bracket fungi, and only a few toadstools, such as *Boletus,* have pores.
 Bracket fungi are usually leathery rather than fleshy and last longer than fleshy toadstools. Most survive for only one season but some such as *Ganoderma,* are perennial. Polypores with leathery or woody fruit bodies may be placed in the order Aphyllophorales.

Economic importance of fungi

Fungi can cause both economic gain and economic loss

Fungi causing economic gain

The most useful fungi are those that are saprophytic in the soil and which, with bacteria, break down dead remains of plants and animals. This is necessary to prevent living organisms from being swamped in dead material, and to enable mineral ions locked up in dead matter to be released and recycled. Without these fungi, normal life and farming could not proceed.

Many fungi are used in biosynthesis, the making of a chemical using a living organism rather than a chemical process *in vitro*. Yeast is one of the best known. It has long been used in leavening bread and in the production of alcohol. Yeast will make dough rise because bubbles of carbon dioxide, which are produced in respiration, are caught in the dough. The production of alcohol requires yeast to respire under anaerobic conditions, and here alcohol and carbon dioxide are formed. When the alcohol content reaches 14% it inhibits yeast action, but alcohol is made more concentrated by distillation.

More recently yeast has been used as a source of B vitamins in products such as Yeast-vite and Marmite, and as a source of protein. Yeast will grow on glucose with nitrates added and will make amino acids and so proteins for its growth and reproduction. The dried yeast, rich in protein, can be used as an animal feed or textured for human use. Protein made in this way using micro-organisms is known as single-cell protein. Today bacteria are being widely used in biosynthesis. The industry growing up from such use is called biotechnology or cell technology, but fungi are still used in many processes.

The green mould, *Penicillium*, is well known as the agent for the biosynthesis of penicillin, the first antibiotic which was discovered by Sir Alexander Fleming in 1928 and published in 1929. The discovery is often called a "happy accident". Fleming was working with petri dishes of bacteria, one of which was accidentally contaminated with *Penicillium*, and he noticed that no bacteria were growing near the fungus. He isolated the mould, *Penicillium notatum,* and found that it would kill Gram positive bacteria, The production of large quantities of penicillin and its use in the treatment of disease came much later, given added impetus by World War II. Other species of *Penicillium* were used, the first being *P. chrysogenum*, which could be grown in deep vats and so produced more penicillin than *P. notatum*, which can only be grown on the surface of a substrate. At first, *P. chrysogenum* was lower yielding than *P. notatum*, but exposure to ultra-violet and X-rays resulted in the development of a new strain with a higher yield.

Other moulds are useful in biosynthesis. For example, *Aspergillus* is used in the production of citric acid (used in soft drinks), itaconic acid (used to make plastics), gluconic acid (used as its calcium salt as a calcium supplement) and various other organic acids.

Penicillium moulds are also used in ripening cheeses. All cheese is hardened with rennet, and in many common cheeses the process stops there and fungi are not used at all. But the blue cheeses, such as Roquefort and Stilton, have spores of *P. roqueforti* inoculated into them. The conidia produce the blue colour, and products of metabolism of the fungus give the strong taste. Soft cheeses with a furry coat like Camembert and Brie have another species, *P. camemberti,* growing on the outside. The mycelium forms the furry rind, and the softness of the cheese inside is due to its fungal digestion.

Certain trees have an association with fungi called mycorrhizae (or mycorrhizas). A mycorrhiza is a union between a tree root and fungal hyphae. Many fungi produce fruit bodies only under a particular type of tree because they have mycorrhizae on its roots, for example, *Boletus* is found under birch and pine. Many mycorrhizae are ectotrophic, i.e. the fungal mycelium encases the root but does not penetrate it. This relationship is thought to be symbiotic; the fungus takes food from the root, but enables the root to absorb minerals from the soil more effectively than can be done by root hairs. When conifers are planted on poor soils they are often infected with spores of ectotrophic mycorrhizal fungi. Some mycorrhizae are endotrophic and penetrate the root and actually enter the root cells. This is the situation of mycorrhizae on orchid roots. Many orchid seeds will not germinate without their mycorrhizae, which has led to problems with their cultivation (now overcome with modern infection techniques and meristem culture). The relationship between the orchid and the fungus is not really understood. Some orchids, such as the bird's nest orchid, lack chlorophyll and entirely dependent on their mycorrhizae. The situation seems to be that the fungus is saprophytic on organic material in the soil and the orchid is parasitic on the fungus.

Many larger fungal fruit bodies are edible. Usually these are Basidiomycetes, but a few, such as truffles and morels, are Ascomycetes. In Britain only the cultivated mushroom, grown under closely controlled conditions, is widely on sale, but in Europe far more species are sold in markets as well as being picked wild, for example the cep, chanterelle and oyster fungus. Truffles are the fruit bodies of Ascomycetes. They are very expensive for two reasons; firstly because they form their fruit bodies underground, and secondly because they have mycorrhizal associations with certain trees, so they cannot be cultivated as an underground crop and cannot easily be found. They are hunted with dogs or pigs. Truffles do grow in Britain but are seldom collected here. The most famous truffle areas are the Dordogne region of France and the Piedmont district of northern Italy.

Fungi causing economic loss

These include fungi which are pathogenic (i.e. parasitic and causing disease to plants, animals and man), those which destroy man's stored products, and those which are poisonous.

Fungi are important plant pathogens. One of the best known in Britain is the potato blight, *Phytophthora infestans*, the fungus responsible for the Irish potato famine in the 1840s. Potatoes grow better than cereals in Ireland, and the Irish peasants of the nineteenth century were almost entirely dependent on them. In 1844 and 1845 the crop and the stored potatoes were completely destroyed by blight. Many people died or emigrated to mainland Britain or the U.S.A.

On continental Europe, North America and India, rusts (Uredinales) and smuts (Ustilaginales) which grow on ears of grain are very damaging. In the past, ergot of rye, *Claviceps purpurea*, has been dangerous, causing the disease ergotism, with symptoms of hallucinations, gangrene and death.

There are a number of mildew diseases, including grape mildew, the downy mildews (*Peronospora*) on crucifers like cabbage, mustard, etc., and the powdery mildews (Erysiphales) which form a white powder on plants.

Seeds may be affected by damping off disease, *Pythium*, and fruits by rots such as *Monilia*, the rot of apples.

Dutch elm disease is caused by a fungus, *Ceratocystis ulmi,* which is transmitted by a beetle. Other fungi transmitted by insects cause blights, galls, and staining of cotton lint.

An important pathogen of trees is the honey fungus, *Armillaria mellea,* which looks like a woodland toadstool, but whose mycelium forms thick strands called rhizomorphs that attack living roots. When it reaches the base of the tree trunk, it can ring the tree, killing it, and then lives on the dead body of its victim as a saprophyte.

Fungi are less important pathogens of man and animals, but ringworm and athlete's foot are examples of fungal diseases. Spores of *Aspergillus* in mouldy hay can cause the respiratory disease aspergillosis, which is more important in domestic animals, but is sometimes found in farm workers, when it is called "farmers' lung".

Saprophytic fungi which destroy man's stored products are called moulds, although the term is used more generally for any saprophytic fungus causing decay. *Penicillium* is the green mould on bread, jams, cheese and citrus fruit. Black moulds may be *Mucor, Rhizopus,* or *Aspergillus,* and many other genera are also found.

In houses, dry rot is caused by *Merulinus (Serpula) lachrymans.* This is a bracket fungus whose hyphae grow on wood, digesting its cellulose. It does not destroy lignin, which remains as a film of powder and falls to dust when touched. The orange and white, bracket-like fruit bodies are seen on walls, stairs, etc. in a bad attack of the fungus.

Many fungi are poisonous. The most dangerous in Britain is the death cap, *Amanita phalloides.* This looks a little like a mushroom, but has white gills and is most often mistaken for a mushroom in its button stage when the gills cannot be seen. It is particularly dangerous because symptoms do not appear until the poison has been absorbed into the bloodstream, when it is very difficult to control, and has been responsible for most cases of death by fungal poisoning. Its relative, the fly agaric, *Amanita muscaria,* is also poisonous, but has been used as a hallucinogen in the religious and social ceremonies of some cultures. The smaller, delicate *Psilocybe* is also hallucinogenic, and has been used in similar ways.

Division Lichenes (lichens)

Characteristics

Lichens are composite plants, each an association between an alga and a fungus forming a single thallus. The relationship is usually considered to be symbiotic, but is sometimes said to be parasitic by the fungus because the alga can live independently without the fungus, but this cannot survive without the alga.

The fungal partner is called the mycobiont and is usually an ascomycete, although occasionally a basidiomycete. It has a typical fruit body and fungal reproduction.

The algal partner is called the phycobiont and may be a green alga, especially *Trebouxia* or sometimes *Pleurococcus*, or a blue-green alga, e.g. *Nostoc*. The algal cells may be in a special layer in the thallus, the heteromerous condition, or scattered throughout the thallus, the homoiomerous condition.

The resulting lichen looks unlike either partner and can live in habitats in which neither could survive by itself, e.g. bare rock surfaces. The fungus enables the plant to live in dry places, and the alga provides food by photosynthesis for both partners. The fungus absorbs water and minerals from the air and rain very effectively. This means that lichens also absorb pollutants, which is the reason that they are very susceptible to air pollution, and can be used to measure it.

There are three groups of lichens; these comprise two classes **Ascolichenes** and **Basidiolichenes,** and a group called **Lichenes Imperfecti** in which the fruit body is not known. In some classifications, lichens are considered to be in the division Fungi, and are classified as Ascomycetes, Basidiomycetes, or Fungi Imperfecti.

Class Ascolichenes (ascolichens)

The fungus is an ascomycete. This group includes most lichens, e.g. *Cladonia*.

Class Basidiolichenes (basidiolichens)

The fungus is a basidiomycete. This is a much smaller group, e.g. *Omphalina*.

There are three types of lichen habit, crustose, foliose and fruticose.
Crustose (crustaceous) lichens are encrusting, flattened structures, especially on rocks, e.g. *Lecanora, Physcia, Xanthoria*.
Foliose (leafy) lichens look somewhat like a flat, lobed leaf, e.g. *Cladonia, Parmelia, Peltigera. Cladonia* has its fruit bodies (apothecia) borne at the top of long, funnel-like tubes called podetia.
Fruticose (shrubby) lichens are more upright and have many branches arising from a single attachment, sometimes hanging down from trees, e.g *Usnea*.

Economic importance of lichens

Many lichens yield dyes. The Scottish woollen industry once depended on lichen dyes, especially in the Hebrides, where Harris tweed cloth was originally dyed with lichens. Species of the lichen *Roccella* yield litmus dye.
   Some lichens are used as animal food, e.g. reindeer ''moss'', *Cladonia rangiferina*, is a lichen on which the reindeer of Lapland depend. Until recently, the reindeer followed the lichen, and the Laplanders followed the reindeer.

Division	Lichenes
Class	Ascolichenes
Genus	Xanthoria

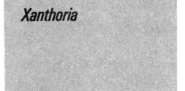

Xanthoria

Thallus with apothecia, surface view

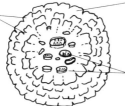

growth occurs by increase around the margins; growth is very slow, and a thallus may survive for many years

vegetative thallus of algal cells and fungal hyphae, coloured with orange pigments, the products of lichen metabolism

apothecia (fruit bodies in which asci develop), appear as little saucers

V.S. through apothecium and thallus

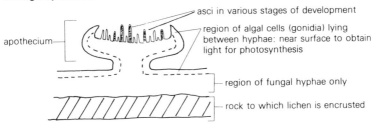

apothecium

asci in various stages of development

region of algal cells (gonidia) lying between hyphae: near surface to obtain light for photosynthesis

region of fungal hyphae only

rock to which lichen is encrusted

Soredium for vegetative reproduction

In some species, the thallus contains structures especially concerned with soredia development.

algal cells

fungal hyphae

Habitat: *Xanthoria* is a crustose lichen, encrusting on walls, roofs, etc. It is one of our most pollutant-resistant lichens. Crustose lichens are one of the primary colonisers of bare rock and start the process of breaking it down to soil by release of products of their metabolism.

Sexual reproduction: is with the formation of ascospores in apothecia as in Discomycetes (see *Peziza*, page 38). In some ascolichens, asci are borne in flask-shaped structures called perithecia. After dispersal, the spores must find a suitable algal partner on the substrate on which they land.

Vegetative reproduction: is by small bodies called soredia consisting of a few algal cells surrounded by fungal filaments. They form a grey, dusty powder which is easily removed from the lichen and gives rise to new plants.

Division Bryophyta

Characteristics

Land plants (usually) with clearly visible alternation of generations (see page 145), both generations being equally conspicuous and almost equally long-lived, although in some genera the sporophyte is only present for a short time.

The gametophyte comprises the leafy plant or thallus. It is the independent photosynthetic generation, but is not well adapted to life on land, and needs to be kept moist if growth is to occur. It usually has no cuticle to conserve water, has no roots to penetrate deep into the soil for water, and no xylem for conduction of water and to give strength. It must be low growing, since it has little strengthening tissue, and to allow for sexual reproduction to occur with swimming spermatozoids.

The sporophyte (which is also called the sporogonium) comprises the stalk and capsule, and depends on the gametophyte throughout its life, but it is better adapted to life on land, as it sticks up into the drier air.

Free water is necessary for sexual reproduction, and the male gamete swims to the female gamete by chemotaxis, so sexual reproduction can only occur in wet conditions.

Gametes are borne in sex organs called antheridia and archegonia and are produced by mitosis. Only the male gamete is released, and fertilisation occurs inside the archegonium.

Dry conditions are necessary for spore dispersal.

Spores are borne in a capsule and are produced by meiosis. Only one type of spore is produced (homosporous).

Summary classification of the division

Division Bryophyta (bryophytes)
  Class Bryopsida (Musci, mosses)
    Subclass Sphagnidae
    Subclass Andreaeidae
    Subclass Bryidae
  Class Hepaticopsida (Hepaticae, liverworts)
    Subclass Jungermanniae
    Subclass Marchantiae
  Class Anthocerotopsida (horned liverworts, hornworts)

Notes

Bryophytes appear in the fossil record at about the same time as pteridophytes. Their life cycles suggest that they are ancestral to pteridophytes because they rely heavily on wet conditions since the gametophyte dries out easily. Pteridophytes, with their shorter gametophyte stages and well adapted sporophyte are better suited to life on land, and so are thought to be evolutionarily more advanced. But the structure of the capsule of mosses, which possesses stomata, suggests that bryophytes may have evolved from some degenerate group of pteridophytes. Another theory is that a group of green algae was the common ancestor for both bryophytes and pteridophytes which then evolved along different routes, and that the bryophytes did not give rise to any other group.

Characteristics of classes

Class Bryopsida (Musci, mosses)

1 Always leafy, never thalloid, and leaves are never two-ranked or lobed; no underleaves; may be erected or creeping.
2 Spores give rise to a protonema which develops into the gametophyte.
3 Rhizoids are multicellular.
4 Archegonium may form cap (calyptra) on top of the capsule.
5 Seta lengthens slowly, as capsule develops; seta often strong and coloured.
6 Lower part of capsule may be photosynthetic with stomata.
7 Dehiscence of capsule is by peristome teeth.

Subclass Sphagnidae (peat or bog mosses)
One genus, *Sphagnum*, which contributes to sphagnum peat.
Subclass Andreaeidae (granite mosses)
Two genera, *Andreaea* and *Neuroloma*, found on granite outcrops. Sometimes another genus, *Acroschisma*, is recognised.
Subclass Bryidae (true mosses)
The largest group of mosses, including all other common mosses, e.g. *Bryum, Funaria, Mnium, Polytrichum, Tortula*.

Class Hepaticopsida (Hepaticae, liverworts)

1 May be thalloid or leafy; if they are leafy, the leaves are lobed and very clearly borne in two ranks with underleaves called amphigastria.
2 Spores develop directly into the gametophyte with no protonema or with a very reduced protonema.
3 Rhizoids are unicellular.
4 Archegonium does not form a cap that is carried up on top of the capsule.
5 Seta lengthens rapidly after development of capsule to full size; seta often weak and colourless.
6 Lower part of capsule is not photosynthetic and does not have stomata.
7 Dehiscence of capsule is by elaters.

Subclass Jungermanniae
1 Little differentiation of thallus.
2 Sex organs are not on stalks.
e.g. *Pellia* (thalloid), *Lophocolea* (leafy).

Subclass Marchantiae
1 Much differentiation of thallus with air pores, etc.
2 Sex organs are borne on stalks.
e.g. *Marchantia* (thalloid), *Lunularia* (thalloid).

Class Anthocerotopsida (horned liverworts, hornworts)

These are somewhat different from true liverworts, They are always thalloid, the capsule is needle-shaped and contains stomata, and there is one single large chloroplast per cell.
e.g. *Anthoceros*.

Tortula muralis	Division	Bryophyta
	Class	Bryopsida
	Subclass	Bryidae
	Order	Pottiales
	Genus	Tortula

Entire plant with young capsule

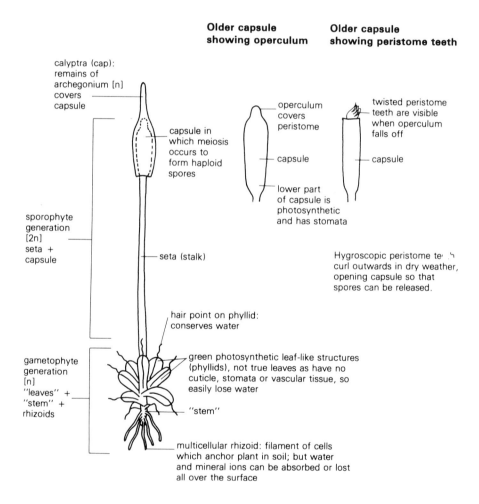

Older capsule showing operculum

Older capsule showing peristome teeth

calyptra (cap): remains of archegonium [n] covers capsule

operculum covers peristome

twisted peristome teeth are visible when operculum falls off

capsule in which meiosis occurs to form haploid spores

capsule

capsule

sporophyte generation [2n] seta + capsule

lower part of capsule is photosynthetic and has stomata

seta (stalk)

Hygroscopic peristome te h curl outwards in dry weather, opening capsule so that spores can be released.

hair point on phyllid: conserves water

gametophyte generation [n] "leaves" + "stem" + rhizoids

green photosynthetic leaf-like structures (phyllids), not true leaves as have no cuticle, stomata or vascular tissue, so easily lose water

"stem"

multicellular rhizoid: filament of cells which anchor plant in soil; but water and mineral ions can be absorbed or lost all over the surface

Habitat: common on walls, even in towns, growing in a mat, not dense cushions.
Notes: all stages are often present in the same mat. When the calyptra falls off it exposes the operculum; this is then shed, exposing the peristome teeth. These are twisted in *Tortula*, but most mosses have an untwisted peristome. Other common and similar mosses on walls include *Grimmia*, which grows in cushions, and *Bryum capillare* which has pendulous capsules.

Life cycle of a moss

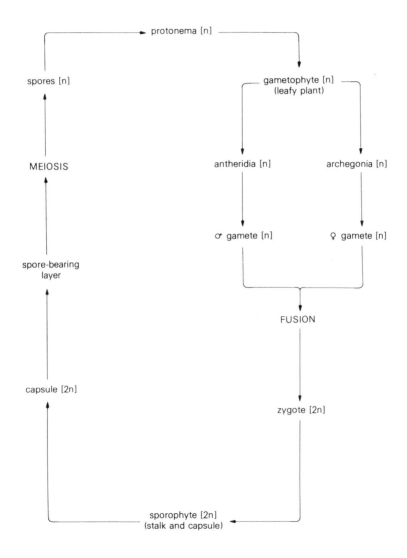

The genus *Tortula* is illustrated because it is a very common moss, but the life cycle described on the next page is that of *Funaria*, because *Funaria* has been studied in detail. *Funaria* is less common than *Tortula* and grows mainly on burnt ground, but is also found on pots in greenhouses. It does not have hair points on its leaves and its peristome is untwisted.

Funaria hygrometrica	*Division*	Bryophyta
	Class	Bryopsida
	Subclass	Bryidae
	Order	Funariales
life cycle of a moss	*Genus*	*Funaria*

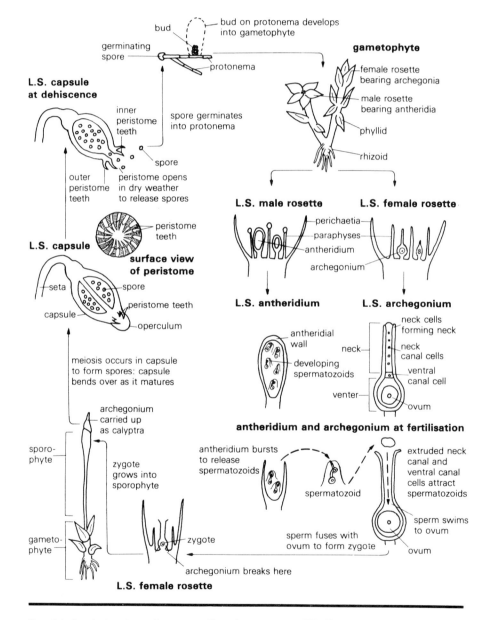

L.S. capsule at dehiscence

L.S. capsule

surface view of peristome

gametophyte

L.S. male rosette

L.S. female rosette

L.S. antheridium

L.S. archegonium

antheridium and archegonium at fertilisation

L.S. female rosette

For details of structure of a moss, *Tortula*, see page 50. For explanation, see page 53.

Life cycle of a moss: *Funaria*

Mosses show alternation of generations in which both the gametophyte and sporophyte are equally conspicuous and both contribute to the typical moss plant. The concept of alternation of generations is explained on page 145, and diagrams of moss structure and life cycle are shown on pages 50 to 52. The gametophyte is the independent generation and the sporophyte depends on it for the whole of its existence, although there is some photosynthesis in the lower part of the capsule, which also has stomata. The gametophyte is not very well adapted to life on land and usually dries out easily, but unlike liverworts, many mosses can survive the drying process, and although they can only grow and reproduce in wet conditions, many can live in much drier places than liverworts.

The sporophyte produces haploid asexual spores by meiosis in the capsule in a special, spore-bearing layer. When the capsule is mature, the calyptra (cap) falls off, exposing the operculum. Then large annulus cells around the edge of the operculum absorb water and force the operculum off, exposing the peristome teeth. There are two rows of these in *Funaria,* with sixteen teeth in each row. Spore dispersal occurs in dry conditions. The peristome teeth are hygroscopic, i.e. sensitive to changes in humidity. They keep the capsule closed by curling inwards when conditions are wet, but the outer teeth curl back and the inner teeth part when the weather is dry, so that the spores can be shaken out. It may be that the inner teeth catch against the base of the outer ones and help to flick the spores out. The spores travel in air currents and land on various substrata.

If a spore lands on a suitable damp place, it germinates to produce a filamentous protonema, similar to a filamentous green alga. Buds appear on the protonema and each grows into a gametophyte. This is vegetative reproduction. Some mosses also reproduce vegetatively by little buds called gemmae on the rhizoids or stem.

The gametophyte produces sex organs and gametes. The sex organs are borne at the tip of leaf-lined rosettes, the leaves of which are called perichaetia. Some mosses are dioecious, but both *Funaria* and *Tortula* are monoecious. Since the gametophyte is haploid, the sex organs and gametes are produced by mitosis. The male sex organs are club-shaped structures called antheridia, and the male gametes are called spermatozoids. The female sex organs are flask-shaped structures called archegonia, each consisting of a neck containing neck canal cells and a ventral canal cell, and a venter containing the female gamete (ovum).

Sexual reproduction occurs in wet conditions. Hair-like paraphyses in both rosettes help to hold water. In the neck of the archegonium, the neck canal cells and ventral canal cell absorb water and become mucilaginous, swelling and forcing open the cap cells at the top of the neck. The mucilage oozes out of the neck, collecting in a pool; it contains a chemical, sucrose in mosses, which attracts the spermatozoids to the archegonia. Meanwhile the antheridia absorb water and the tops burst open, liberating the flagellated spermatozoids, which swim to the archegonia by chemotaxis in the surface film of water. One swims down the neck of the archegonium and fuses with the ovum to form a zygote. Cross fertilisation usually occurs as antheridia and archegonia mature at different times.

The zygote divides by mitosis to form the diploid sporophyte generation of seta and capsule. The seta elongates gradually as the capsule develops, and is carried up on top of the capsule as the cap or calyptra, which falls off as the capsule matures.

Pellia epiphylla	*Division*	**Bryophyta**
	Class	**Hepaticopsida**
	Subclass	**Jungermanniae**
	Order	**Metzgeriales**
	Genus	*Pellia*

External structure   **Dehiscing capsule**

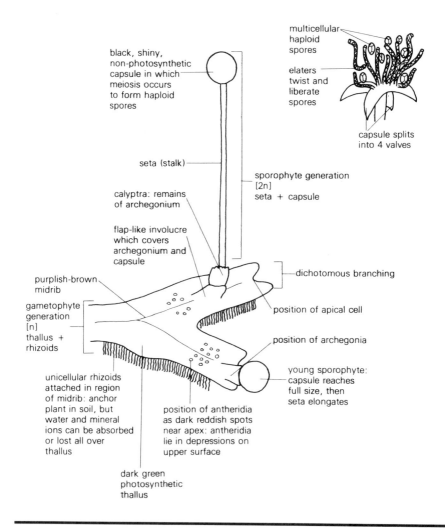

black, shiny, non-photosynthetic capsule in which meiosis occurs to form haploid spores

multicellular haploid spores

elaters twist and liberate spores

capsule splits into 4 valves

seta (stalk)

sporophyte generation [2n]
seta + capsule

calyptra: remains of archegonium

flap-like involucre which covers archegonium and capsule

purplish-brown midrib

gametophyte generation [n]
thallus + rhizoids

dichotomous branching

position of apical cell

position of archegonia

young sporophyte: capsule reaches full size, then seta elongates

unicellular rhizoids attached in region of midrib: anchor plant in soil, but water and mineral ions can be absorbed or lost all over thallus

position of antheridia as dark reddish spots near apex: antheridia lie in depressions on upper surface

dark green photosynthetic thallus

Habitat: common in moist banks and ditches near streams.
Notes: there are two other very common thalloid liverworts, *Marchantia* and *Lunularia*. Both differ from *Pellia* in having pores on the surface of the thallus, and in having gemmae cups; these are round in *Marchantia* but half-moon-shaped in *Lunularia*. *Marchantia* bears antheridia and archegonia on stalks (see pages 58, 59).

Life cycle of a liverwort

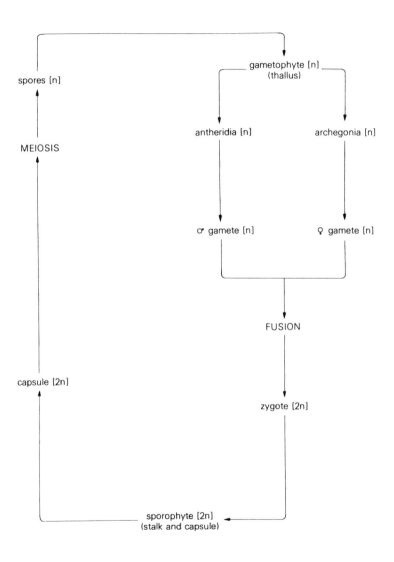

Pellia epiphylla	*Division*	Bryophyta
	Class	Hepaticopsida
	Subclass	Jungermanniae
life cycle of a	*Order*	Metzgeriales
thalloid liverwort	*Genus*	Pellia

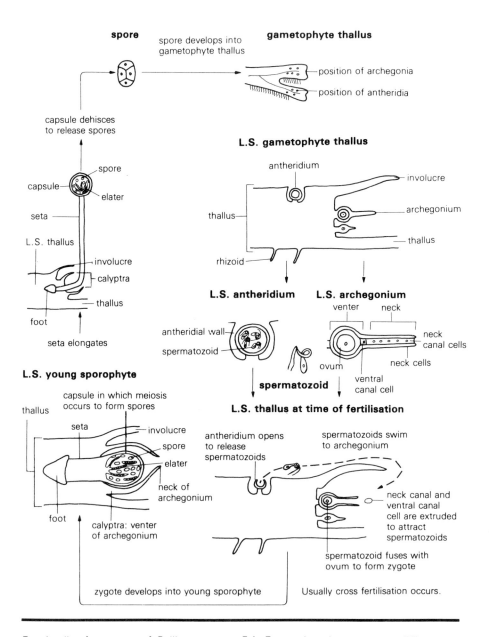

For details of structure of *Pellia*, see page 54. For explanation, see page 57.

Life cycle of a liverwort: *Pellia*

Liverworts show alternation of generations in which both the gametophyte and sporophyte are equally conspicuous, although the sporophyte may only be present for a short time. The concept of alternation of generations is explained on page 145, and diagrams of liverwort structure and life cycle are shown on pages 54 to 56. The gametophyte is the independent generation, and the sporophyte depends on it for the whole of its existence. The gametophyte is not very well adapted to life on land and can dry out easily, so liverworts can only survive in damp places.

The sporophyte produces haploid asexual spores by meiosis inside the capsule. These spores then undergo mitosis and acquire chloroplasts before they are dispersed. Spore dispersal occurs in dry conditions. As the capsule loses water, the wall shrinks and splits into four valves. Inside, there are long cells called elaters each with a double helix of hygroscopic tissue; as they dry, the elaters twist, shooting out spores which are dispersed in air currents. In the centre of the capsule is a tuft of elater-like cells joined at the base of the capsule and known as the elaterophore. As this dries, its cells flick from side to side assisting spore dispersal. After dehiscence, the elaterophore persists as a tuft in the capsule.

If a spore lands on a suitable damp place, it germinates to produce the gametophyte thallus. In some liverworts, but not in *Pellia*, little cups known as gemmae cups are seen on the surface of the thallus and produce multicellular buds called gemmae which are dispersed in rain splash and grow into new gametophytes. This is vegetative reproduction. Gemmae are illustrated on page 59.

The gametophyte produces sex organs and gametes. Usually both sex organs are borne on the same thallus (monoecious). Since the gametophyte is haploid,these structures are produced by mitosis. The male sex organs are stalked club-shaped structures called antheridia, and the male gametes are spermatozoids. The female sex organs are flask-shaped structures called archegonia, each consisting of a neck containing neck canal cells and a ventral canal cell, and a venter containing the female gamete (ovum). Archegonia develop under a flap-like involucre.

Sexual reproduction occurs in wet conditions. In the neck of the archegonium the neck canal cells and ventral canal cell absorb water and become mucilaginous, swelling and forcing open the cap cells at the top of the neck. The mucilage oozes out of the neck, collecting in a pool; it contains a chemical, a protein in liverworts, which attracts the spermatozoids to the archegonia. Meanwhile the antheridia absorb water and the top bursts, liberating the flagellated spermatozoids which swim to the archegonia by chemotaxis in the surface film of water. One swims down the neck of the archegonium and fuses with the ovum to form a zygote. Cross fertilisation usually occurs, as antheridia and archegonia mature at different times.

The zygote divides by mitosis to form the sporophyte generation of seta, capsule, and foot which absorbs food for the sporophyte from the gametophyte. The capsule matures first and reaches mature size on the thallus, protected by the venter of the archegonium which grows to keep pace, and is now called the calyptra. The sporophyte remains at this stage over winter. In early spring, the seta elongates rapidly, breaking through the calyptra, so it is not carried up on top of the capsule as a cap, which happens in mosses.

Marchantia polymorpha	Division	Bryophyta
	Class	Hepaticopsida
	Subclass	Marchantiae
	Order	Marchantiales
	Genus	*Marchantia*

Stages of life cycle in *Marchantia*

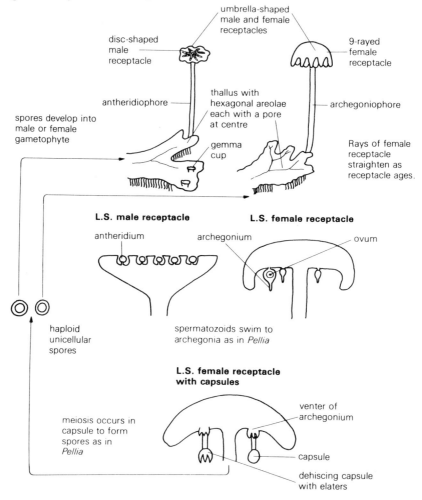

Habitat: in gardens, etc. on moist brickwork and flower pots; quite common.
Notes: the life cycle is similar to that of *Pellia*, but sex organs are on stalks and capsules are formed below the archegonial head. The sex of the gametophytes is controlled by an XY mechanism. The sporophyte chromosome complement is 2n + XY, and the gametophytes are n + X (female) and n + Y (male), produced in equal numbers.

		Division	Bryophyta		
		Class	Hepaticopsida		
Subclass	Marchantiae			*Subclass*	Jungermanniae
Order	Marchantiales			*Order*	Jungermanniales
Genus	Lunularia			*Genus*	Lophocolea

External features of *Lunularia cruciata*, a thalloid liverwort

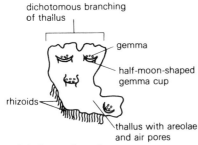

dichotomous branching of thallus

gemma

half-moon-shaped gemma cup

rhizoids

thallus with areolae and air pores

Part of thallus, enlarged

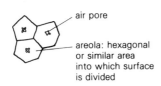

air pore

areola: hexagonal or similar area into which surface is divided

Gemmae dispersal in typical cup-shaped gemmae cup

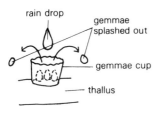

rain drop

gemmae splashed out

gemmae cup

thallus

Gemmae are small, non-sexual vegetative reproductive bodies dislodged by swelling of mucilage in gemmae cup, and dispersed in rain splash.

Most gemmae cups are cup-shaped. The shape in *Lunularia* is unusual and diagnostic.

External features of *Lophocolea bidentata* a leafy liverwort

Upper surface

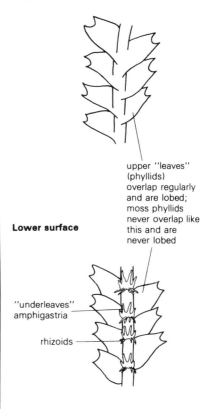

upper "leaves" (phyllids) overlap regularly and are lobed; moss phyllids never overlap like this and are never lobed

Lower surface

"underleaves" amphigastria

rhizoids

Habitat: *Lunularia* is found in gardens, etc., on walls, paths, lawns and on flower pots; one of the commonest garden liverworts.

Habitat: *Lophocolea* is a common liverwort in grass, and is found in wet lawns, pastures, and grassy banks.

Division Pteridophyta

Characteristics

Land plants with alternation of generations (see page 145), the sporophyte being the conspicuous and dominant generation and the gametophyte small and inconspicuous.

The gametophyte (prothallus) is the independent generation, but is not well adapted to life on land. It is a small flat plate of photosynthetic cells with no cuticle to prevent drying out. Rhizoids anchor it to the soil, but water can be taken up or lost all over the surface.

The sporophyte is the main plant, the fern, horsetail, club moss, or whisk fern. It develops from the gametophyte and depends upon it for the early part of its life, but is much better adapted to life on land. The leaves have a cuticle to conserve water and stomata for entry and exit of gases; roots penetrate deep into the soil to absorb water at depth, and xylem is present to transport water through the plant and to give strength.

Free water is necessary for sexual reproduction, and the male gamete swims to the female by chemotaxis, so sexual reproduction can only occur in wet conditions.

Gametes are borne in sex organs called antheridia and archegonia which are sometimes very reduced; gametes are produced by mitosis. Only the male gamete is released and fertilisation occurs inside the archegonium.

Dry conditions are necessary for spore dispersal.

Spores are borne in sporangia and are produced by meiosis. One type of spore may be produced (homosporous) or two types (heterosporous).

Summary classification of the division

Division Pteridophyta (pteridophytes)
  Class Filicopsida (ferns)
  Class Sphenopsida (horsetails)
  Class Lycopsida (club mosses)
  Class Psilophytopsida ⎤
  Class Psilotopsida   ⎦— Class Psilopsida (whisk ferns)

Notes

Pteridophytes are the earliest land plants to be found in the fossil record, with some forms in Cambrian, Silurian and Devonian rocks. They formed the dominant vegetation of the Carboniferous period when they reached tree size and contributed to the Coal Measures.

The life cycles of pteridophytes are intermediate between those of bryophytes and spermatophytes. Although the sporophyte is well adapted to life on land, the gametophyte needs water for its survival and for sexual reproduction to occur, so they tend to live in wet places. The gametophyte must be small and low growing since sexual reproduction requires swimming sperm. Spermatozoids could not swim to the top of a tall tree, or even to the top of a fern frond.

There is an extinct group of plants called pteridosperms or seed ferns, which have large fern-like fronds but cones like gymnosperms. They are considered to be a group of gymnosperms, and are discussed as the order Pteridospermales on page 77.

Characteristics of classes

Class Filicopsida (ferns)
1 Leaves are large and called fronds.
2 Spores are borne in sporangia on the underside of leaves, usually ordinary foliage leaves.
3 Usually only one kind of spore is produced (homosporous) and only one kind of gametophyte which bears both antheridia and archegonia. A few ferns, including a small group called water ferns, are heterosporous.
e.g. *Dryopteris* (male fern), *Pteridium* (bracken), *Phyllitis* (hart's tongue), *Adiantum* (maidenhair fern), *Osmunda* (royal fern), *Ophioglossum* (adder's tongue), *Botrychium* (moonwort). The last three ferns are unusual in that their sporangia are borne on special branches rather than on ordinary foliage leaves. Most modern ferns are in the order Filicales.
　　Ferns are thought to be the group which gave rise to the spermatophytes.
　　The Filicopsida is also sometimes called Pteropsida or Polypodiopsida.

Class Sphenopsida (horsetails)
1 Leaves are small and wedge-shaped and arranged in whorls.
2 Spores are borne in sporangia on sporangiophores arranged in terminal cones.
3 Only one kind of spore is produced (homosporous), but spores often develop into separate male and female gametophytes, the male bearing antheridia and the female bearing archegonia.
e.g. *Equisetum*, which is the only living genus, and is placed with several fossil genera in the order Equisetales. There are several orders of fossil horsetails, including the Calamitales, in which *Calamites* reached tree size.

Class Lycopsida (club mosses)
1 Leaves are small, and often arranged in dorsal and ventral series.
2 Spores are borne in sporangia in the axils of sporophylls on lateral cones.
3 There may be one kind of spore (homosporous) or two kinds of spore (heterosporous). In heterosporous forms, the spores are small microspores and large megaspores produced in micro- and megasporangia. The microspores develop into male gametophytes which bear antheridia, and the megaspores into female gametophytes which bear archegonia.
e.g. *Lycopodium*, order Lycopodiales, which is homosporous.
　　Selaginella, order Selaginellales, which is heterosporous.
　　Isoetes (quillwort), order Isoetales.
　　Lepidodendron, order Lepidodendrales, a fossil reaching tree size.
　　Aldanophyton, order Protolepidodendrales, the earliest known pteridophyte, a fossil found in Cambrian rocks in Siberia.

Class Psilophytopsida
This is a class of extinct pteridophytes, found mainly in Devonian rocks, and only known from the sporophyte. They grew from rhizomes and their aerial branches were either naked or with small, spirally arranged appendages; homosporous.
e.g. *Cooksonia, Rhynia, Yarravia, Zosterophyllum.*

Class Psilotopsida
This is a class of living pteridophytes very similar to the Psilophytopsida, with scale-like or leaf-like appendages. Two genera, *Psilotum* and *Tmesipteris.*

Dryopteris filix-mas	Division	Pteridophyta
male fern	Class	Filicopsida
	Order	Filicales
	Genus	Dryopteris

External structure shown diagrammatically

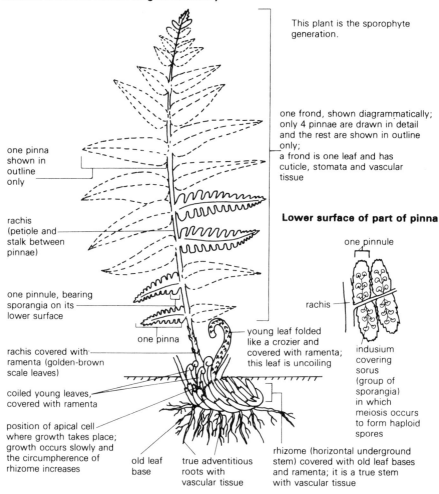

This plant is the sporophyte generation.

one frond, shown diagrammatically; only 4 pinnae are drawn in detail and the rest are shown in outline only; a frond is one leaf and has cuticle, stomata and vascular tissue

one pinna shown in outline only

rachis (petiole and stalk between pinnae)

one pinnule, bearing sporangia on its lower surface

rachis covered with ramenta (golden-brown scale leaves)

coiled young leaves, covered with ramenta

position of apical cell where growth takes place; growth occurs slowly and the circumherence of rhizome increases

one pinna

young leaf folded like a crozier and covered with ramenta; this leaf is uncoiling

old leaf base

true adventitious roots with vascular tissue

rhizome (horizontal underground stem) covered with old leaf bases and ramenta; it is a true stem with vascular tissue

Lower surface of part of pinna

one pinnule

rachis

indusium covering sorus (group of sporangia) in which meiosis occurs to form haploid spores

Habitat: common in woodlands in all parts of Britain. Some species of *Dryopteris* are grown as garden plants, and may also be found with other ferns on walls, since the gametophyte can survive in moist conditions in cracks in mortar.

Notes: the presence or absence of an indusium, its shape, and the position of sori vary in different ferns.

Ferns usually have a thin cuticle on the leaves and are confined to damp places where the gametophyte can survive, but bracken lives on dry commons because it spreads by its rhizome and so avoids the gametophyte stage.

Life cycle of a fern

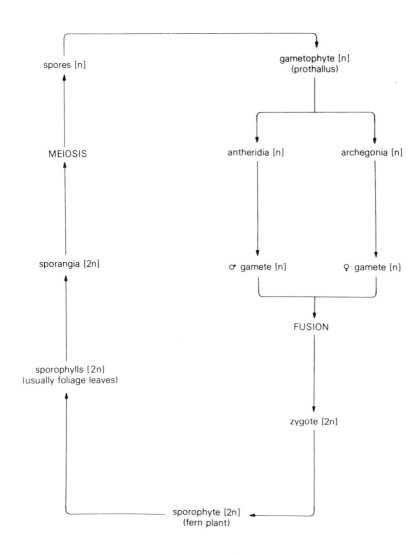

spores [n]

gametophyte [n]
(prothallus)

MEIOSIS

antheridia [n]   archegonia [n]

sporangia [2n]

♂ gamete [n]   ♀ gamete [n]

FUSION

sporophylls [2n]
(usually foliage leaves)

zygote [2n]

sporophyte [2n]
(fern plant)

Dryopteris filix-mas	Division	Pteridophyta
male fern	Class	Filicopsida
	Order	Filicales
	Genus	Dryopteris
life cycle of a fern		

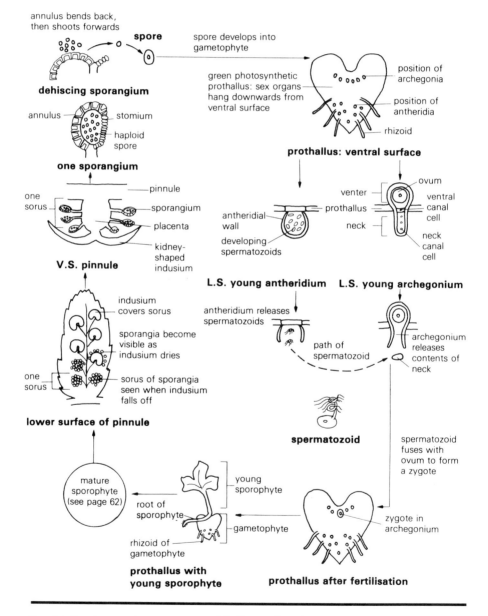

For details of sporophyte, see page 62. For explanation, see page 65.

Life cycle of a fern: *Dryopteris*

Ferns show alternation of generations in which the sporophyte is the most conspicuous generation, forming the fern plant, while the gametophyte is reduced. One type of spore is produced, so ferns are homosporous. The concept of alternation of generations is explained on page 145, and diagrams of fern structure and life cycle are shown on pages 62 to 64. The sporophyte is well adapted to life on land, but although it is the independent generation for most of its life, it depends on the vulnerable gametophyte at the start of its existence. The gametophyte is a small structure called a prothallus, which is much less well adapted to life on land; this is why most ferns can only live in damp places where the gametophyte can survive.

The sporophyte produces haploid asexual spores by meiosis in sporangia on the underside of foliage leaves. Sporangia are borne in clusters called sori and each is covered by an umbrella-like structure called an indusium. In *Dryopteris* the indusium is kidney-shaped, which is diagnostic of the genus. At maturity, in dry weather the indusium falls off and the sori are seen as brown patches on the leaves. Not all ferns have indusia, and the arrangement of sori is different in different ferns.

Dehiscence of the sporangium occurs in dry conditions. The sporangium wall is in two parts, the annulus whose cell walls are thickened on all but the outer face, and the stomium which is thin-walled. As the annulus cells dry, the outer wall bends inwards, pulling the radial walls of each cell towards each other so that the whole distance around the edge of the annulus shortens, exerting a strain. The sporangium wall breaks at its thinnest part, the stomium, to release the strain, and the annulus curls back on itself carrying the spore mass with it. Many spores fall free at this stage. Drying continues until air bubbles appear in the annulus cells; then the annulus snaps back to its original position, catapulting out the spores, which are dispersed in air currents.

If a spore lands on a suitable damp place, it germinates into the gametophyte generation called the prothallus. This is a heart-shaped plate of photosynthetic cells with no cuticle, stomata or vascular tissue; rhizoids anchor it to the substratum, but water and minerals can be taken up and lost all over the surface, so the prothallus can only survive in damp places. It bears the sex organs and gametes. The sex organs are antheridia and archegonia and the gametes are spermatozoids and ova as in bryophytes, but the sex organs are simpler than in bryophytes, e.g. the archegonia have shorter necks, there are no paraphyses, and the sex organs are not on stalks. Since the gametophyte is already haploid, the gametes are produced by mitosis.

Sexual reproduction occurs in wet conditions. In the neck of the archegonium the neck canal cells and ventral canal cell absorb water and become mucilaginous, forcing open the archegonial neck and oozing out; the mucilage contains a chemical, malic acid in pteridophytes, which attracts the spermatoziods. Meanwhile the antheridia absorb water and the tops burst, liberating the flagellated spermatozoids which swim to the archegonia by chemotaxis in the surface film of water. One swims down the neck of the archegonium and fuses with the ovum to form a zygote. Cross fertilisation usually occurs as antheridia and archegonia mature at different times.

The zygote develops into the diploid sporophyte generation. This is dependent on the gametophyte prothallus for its nutrients at first, but it soon puts down a root, opens its leaves, and starts to photosynthesise and becomes independent.

Equisetum species	Division	Pteridophyta
horsetail	Class	Sphenopsida
	Order	Equisetales
	Genus	*Equisetum*

External features of *Equisetum*

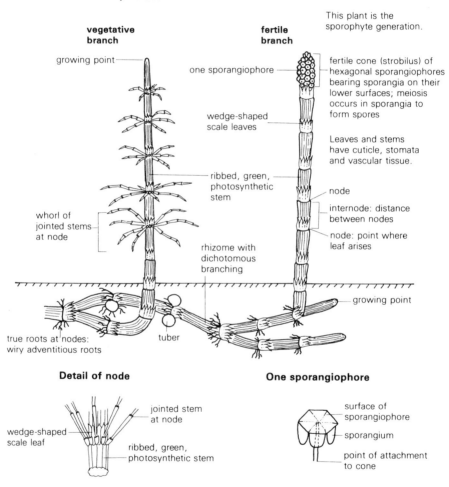

vegetative branch

fertile branch

This plant is the sporophyte generation.

growing point

one sporangiophore

fertile cone (strobilus) of hexagonal sporangiophores bearing sporangia on their lower surfaces; meiosis occurs in sporangia to form spores

wedge-shaped scale leaves

Leaves and stems have cuticle, stomata and vascular tissue.

ribbed, green, photosynthetic stem

node

internode: distance between nodes

node: point where leaf arises

whorl of jointed stems at node

rhizome with dichotomous branching

growing point

true roots at nodes: wiry adventitious roots

tuber

Detail of node

jointed stem at node

wedge-shaped scale leaf

ribbed, green, photosynthetic stem

One sporangiophore

surface of sporangiophore

sporangium

point of attachment to cone

Habitat: *Equisetum arvense* is a common weed of waste places and cultivated land, fields, woods, railway embankments, etc. It spreads vegetatively by its rhizome and tubers and avoids sexual reproduction, so it can live in drier conditions than most horsetails, many of which can only grow in moist places.
Notes: some species have cones on the same branches as those bearing whorls of jointed stems.

The name Sphenopsida comes from the Greek *sphen* (= wedge), referring to the shape of the scale leaves. The class is also known as Articulatae, from the Latin *articulus* (= joint), referring to the jointed branches.

Life cycle of a horsetail

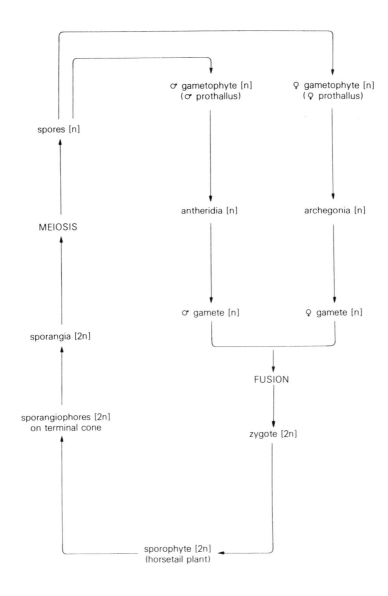

Equisetum species horsetail life cycle of a horsetail	**_Division_** Pteridophyta
	Class Sphenopsida
	Order Equisetales
	Genus _Equisetum_

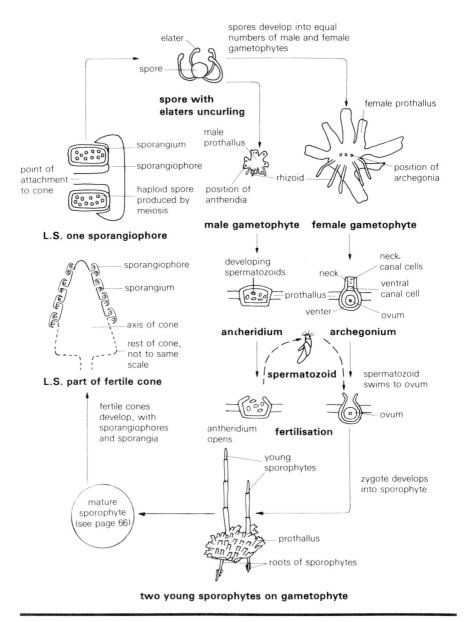

spores develop into equal numbers of male and female gametophytes

elater

spore

spore with elaters uncurling

female prothallus

sporangium

point of attachment to cone

sporangiophore

male prothallus

position of archegonia

rhizoid

haploid spore produced by meiosis

position of antheridia

L.S. one sporangiophore

male gametophyte　**female gametophyte**

sporangiophore

sporangium

developing spermatozoids

neck

neck canal cells

ventral canal cell

prothallus

venter

ovum

axis of cone

antheridium

archegonium

rest of cone, not to same scale

spermatozoid

spermatozoid swims to ovum

L.S. part of fertile cone

ovum

fertile cones develop, with sporangiophores and sporangia

antheridium opens

fertilisation

zygote develops into sporophyte

young sporophytes

mature sporophyte (see page 66)

prothallus

roots of sporophytes

two young sporophytes on gametophyte

For details of sporophyte, see page 66. For explanation, see page 69.

Life cycle of a horsetail: *Equisetum*

Horsetails show alternation of generations in which the sporophyte is the most conspicuous generation, forming the horsetail plant, while the gametophyte is reduced. One type of spore is produced, so horsetails are homosporous. The concept of alternation of generations is explained on page 145, and diagrams of horsetail structure and life cycle are shown on pages 66 to 68. The sporophyte is well adapted to life on land, but although it is the independent generation for most of its life, it depends on the vulnerable gametophyte at the start of its existence. The gametophyte is a small structure called a prothallus, which is much less well adapted to life on land; this is why most horsetails can only live in damp places where the gametophyte can survive. *Equisetum arvense* has overcome the problem by reproducing by its rhizome and tubers and so can live in drier places.

The sporophyte produces haploid asexual spores by meiosis in sporangia which are arranged on the lower surface of sporangiophores borne in terminal cones. The spores, which are large, multicellular and contain chloroplasts, are dispersed in dry conditions. The sporangia open by longitudinal slits. Each spore has an outer wall which splits into four strips; those strips act as elaters, curling when the weather is wet and uncurling when it is dry, so shooting spores out of the sporangia and ensuring that they are only dispersed in dry weather.

If a spore lands on a suitable damp place, it germinates to produce the gametophyte generation, which consists of a prothallus, sex organs and gametes. In some species, the prothallus bears both male and female sex organs, but in most horsetails there are separate male and female prothalli. Both prothalli are lobed structures. Small male prothalli bear the male sex organs, antheridia, sunk into the surface and containing spermatozoids. Larger female prothalli bear the female sex organs, archegonia, projecting from their surface and containing the female gametes, ova. Prothalli are very long lived, and it seems that in some species the large prothallus bears first archegonia, and if these are not fertilised, antheridia, followed by another crop of archegonia.

The structure of the sex organs is similar to those in ferns. Since the prothalli are haploid, all of these structures are produced by mitosis. The prothalli are cushion-like thalli, with plates of photosynthetic cells near the upper surface, and colourless cells below; they have no cuticle, stomata or vascular tissue and are anchored to the soil by rhizoids, but water and minerals can be taken up or lost all over the surface, so the gametophyte easily dries out, and can only survive in damp places.

Sexual reproduction occurs in wet conditions. In the neck of the archegonium the neck canal cells and ventral canal cell absorb water and become mucilaginous, forcing open the archegonial neck and oozing out; the mucilage contains a chemical, malic acid in pteridophytes, which attracts the spermatozoids. Meanwhile the antheridia absorb water and the tops burst, liberating the flagellated spermatozoids which swim to the archegonia by chemotaxis in the surface film of water. One swims down the neck of the archegonium and fuses with the ovum to form a zygote.

The zygote develops into the diploid sporophyte generation. Several sporophytes can develop on one gametophyte, which is unusual. Soon the sporophyte puts down roots, acquires chlorophyll and begins to photosynthesise, becoming the independent generation.

Selaginella species		
Division	Pteridophyta
Class	Lycopsida
Order	Selaginellales
Genus	*Selaginella*

Part of vegetative branch with a fertile cone

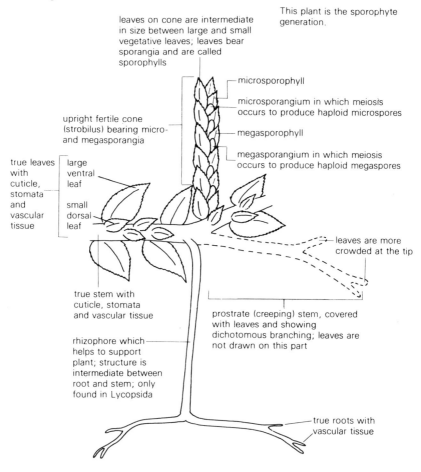

leaves on cone are intermediate in size between large and small vegetative leaves; leaves bear sporangia and are called sporophylls

This plant is the sporophyte generation.

microsporophyll

microsporangium in which meiosis occurs to produce haploid microspores

upright fertile cone (strobilus) bearing micro- and megasporangia

megasporophyll

megasporangium in which meiosis occurs to produce haploid megaspores

true leaves with cuticle, stomata and vascular tissue — large ventral leaf — small dorsal leaf

leaves are more crowded at the tip

true stem with cuticle, stomata and vascular tissue

prostrate (creeping) stem, covered with leaves and showing dichotomous branching; leaves are not drawn on this part

rhizophore which helps to support plant; structure is intermediate between root and stem; only found in Lycopsida

true roots with vascular tissue

Habitat: there is one British species of *Selaginella, S. selaginelloides,* which grows on mountain pastures in northern and western Britain. The African species, *S. kraussiana* is a very common greenhouse plant and is naturalised in some places; it has large and small leaves, while *S. selaginelloides* has only one size of leaf. Not all species have toothed leaves. In the tropics, *Selaginella* has xeromorphic adaptations to survive in dry places, and some are "resurrection plants" which can appear dead but come to life again in water.

Notes: some species grow upright and rhizophores help to support the stems. Usually microsporangia are borne near the top of the cone.

Life cycle of a heterosporous club moss

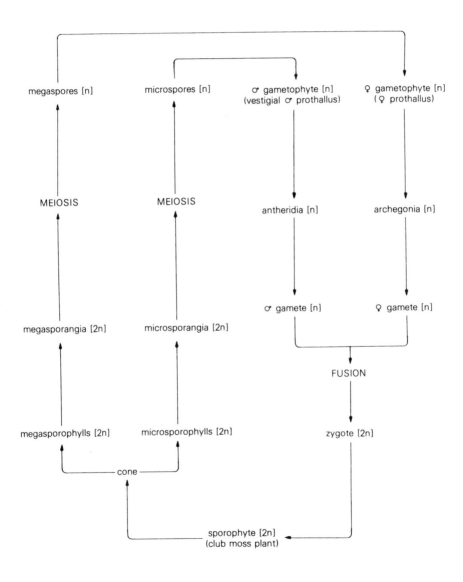

Selaginella species	Division	Pteridophyta
	Class	Lycopsida
	Order	Selaginellales
life cycle of a heterosporous club moss	Genus	Selaginella

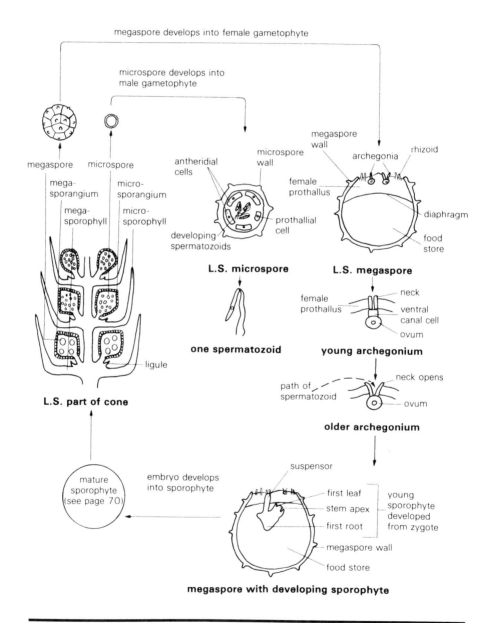

megaspore develops into female gametophyte

microspore develops into male gametophyte

megaspore   microspore

mega-sporangium   micro-sporangium

mega-sporophyll   micro-sporophyll

antheridial cells

microspore wall

megaspore wall   archegonia   rhizoid

female prothallus

developing spermatozoids

prothallial cell

diaphragm

food store

L.S. microspore   **L.S. megaspore**

ligule

one spermatozoid

female prothallus   neck

ventral canal cell

ovum

young archegonium

L.S. part of cone

path of spermatozoid   neck opens

ovum

older archegonium

mature sporophyte (see page 70)

embryo develops into sporophyte

suspensor

first leaf

stem apex

first root

megaspore wall

food store

young sporophyte developed from zygote

megaspore with developing sporophyte

For details of sporophyte, see page 70. For explanation, see page 73.

Life cycle of a heterosporous club moss:
Selaginella

Club mosses show alternation of generations in which the sporophyte is the most conspicuous generation forming the club moss plant, while the gametophyte is very reduced. In *Selaginella* two kinds of spore are produced, so *Selaginella* is heterosporous. The concept of alternation of generations is explained on page 145, and diagrams of *Selaginella* structure and life cycle are shown on pages 70 to 72. The sporophyte is well adapted to life on land and the gametophyte begins to develop on the sporophyte, but the gametophyte is very reduced and water is needed for sexual reproduction, so many *Selaginellas* live only in damp places.

The sporophyte produces haploid asexual spores by meiosis in sporangia on a special vertical branch on which all the leaves (sporophylls) are the same size. Sporangia are borne in the axils of sporophylls. There are two kinds of sporangia. These are microsporangia in which meiosis occurs to form microspores and megasporangia in which meiosis occurs to form megaspores. The sporophylls subtending microsporangia are called microsporophylls and those subtending megasporangia are megasporophylls. The sporophylls and sporangia are the same size, but microspores are much smaller than megaspores. Many microspores are produced, but usually only four megaspores are formed.

The spores develop into the gametophyte generations. Unlike other pteridophytes, development of the gametophytes begins in the spores before they are shed, and nutrients for their development come from the sporophyte. The sporangia dehisce by slit-like openings in dry conditions, and spores may be violently discharged, encouraging cross fertilisation. They germinate into very reduced gametophytes.

The haploid microspore contains the male gametophyte. This is very small and consists of a male prothallus of one prothallial cell, an antheridium which is reduced to a wall, and spermatozoids. The male gametophyte is so small that it does not leave the microspore. Food for its development is stored in the spore and was produced by the parent sporophyte, so the microspore is not really independent.

The haploid megaspore contains the female gametophyte. This begins to develop inside the spore wall before the spores are shed, and food comes from the sporophyte. The megaspore nucleus divides by mitosis to form female prothallus at the top of the spore. Below this are cells which become full of starch produced from the sporophyte. These cells are separated from the female prothallus by a diaphragm, and provide nutrients when the spore is shed. As the female prothallus grows, the spore wall splits. Once on the ground the female prothallus develops rhizoids whose function seems to be to trap water for fertilisation or to trap sperm. The female prothallus does not usually become photosynthetic, and the food used for its development is starch from the sporophyte. Simple archegonia develop on the female prothallus.

Sexual reproduction occurs in wet conditions. It usually happens in the same way as in other pteridophytes, with the liberation of spermatozoids which swim to the archegonium by chemotaxis. One spermatozoid swims down the neck of the archegonium and fuses with the ovum to form a zygote. The zygote develops into the diploid sporophyte generation. At first it depends on the food stores in the megaspore, but it soon begins to photosynthesise and become independent.

In most species of *Selaginella* gametophytes develop on the ground. But in a few species, megaspores remain in megasporangia and produce female gametophytes there. Fertilisation occurs in wet megasporangia by microspores dispersed there and the whole megaspore is shed as a ''seed''.

Lycopodium clavatum	Division	Pteridophyta
	Class	Lycopsida
	Order	Lycopodiales
	Genus	Lycopodium

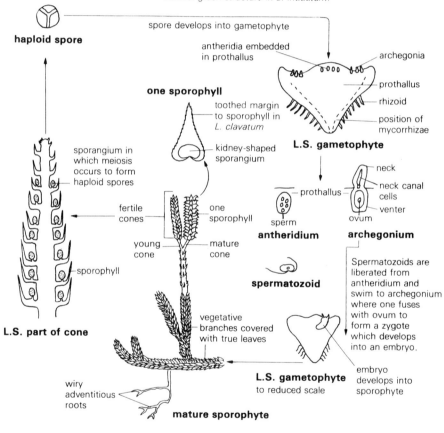

The gametophyte generation in *L. selago* and *L. clavatum* is a subterranean structure which obtains its nutrients from symbiotic mycorrhizae in the prothallus. It is a normal green structure in *L. inudatum*.

haploid spore

spore develops into gametophyte

antheridia embedded in prothallus

archegonia

prothallus

rhizoid

one sporophyll

toothed margin to sporophyll in *L. clavatum*

position of mycorrhizae

L.S. gametophyte

kidney-shaped sporangium

sporangium in which meiosis occurs to form haploid spores

neck

neck canal cells

prothallus

venter

fertile cones

one sporophyll

sperm

ovum

antheridium

archegonium

sporophyll

young cone

mature cone

Spermatozoids are liberated from antheridium and swim to archegonium where one fuses with ovum to form a zygote which develops into an embryo.

spermatozoid

L.S. part of cone

vegetative branches covered with true leaves

embryo develops into sporophyte

L.S. gametophyte to reduced scale

wiry adventitious roots

mature sporophyte

Habitat: five species of *Lycopodium* are native to Britain, found on wet heaths and moorlands of uplands, and there are also many tropical species.

Notes: *Lycopodium* is homosporous; kidney-shaped sporangia are borne in the axils of sporophylls arranged on fertile branches. Meiosis occurs in sporangia to produce haploid spores each of which develops into a prothallus bearing antheridia and archegonia. Fertilisation occurs as in other pteridophytes, and the young embryo begins to develop on the prothallus before becoming independent.

The most common British species are *L. clavatum* (stag's horn moss or wolf's claw) in which the sporophylls are arranged in clear cones, and *L. selago* (fir club moss) in which the fertile branch is similar to the rest of the plant.

Division	Pteridophyta
Class	Psilophytopsida
Order	Psilophytales
Genus	*Rhynia*

Rhynia

Reconstruction of *Rhynia*

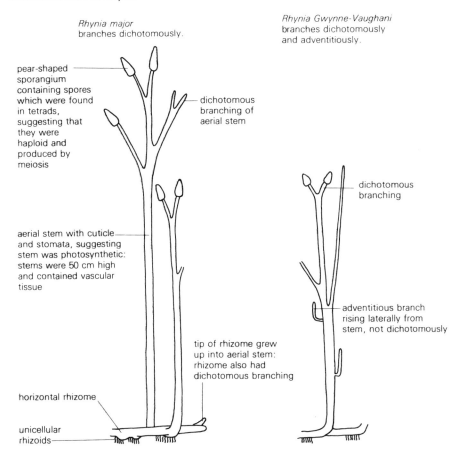

Rhynia major
branches dichotomously.

Rhynia Gwynne-Vaughani
branches dichotomously
and adventitiously.

pear-shaped
sporangium
containing spores
which were found
in tetrads,
suggesting that
they were
haploid and
produced by
meiosis

dichotomous
branching of
aerial stem

dichotomous
branching

aerial stem with cuticle
and stomata, suggesting
stem was photosynthetic:
stems were 50 cm high
and contained vascular
tissue

adventitious branch
rising laterally from
stem, not dichotomously

tip of rhizome grew
up into aerial stem:
rhizome also had
dichotomous branching

horizontal rhizome

unicellular
rhizoids

Rhynia is one of the plants discovered in 1917 in the chert beds at Rhynie in
Scotland. The plants were fossilised in the mid-Devonian period, and are thought to
have grown in a peat bog which became impregnated with silica. Although they
had no leaves, their sporangia and internal vascular tissue suggests that they are
early pteridophytes. It is now thought that fern fronds evolved by enlargement and
flattening of many-branched stems of a *Rhynia*-like plant, and that club mosses and
horsetails evolved from other psilophytes.

Rhynia was once thought to be the earliest land plant, but now other genera of
Psilophytopsida, *Cooksonia* and *Yarravia* have been found in Silurian rocks, and a
fossil club moss, *Aldanophyton*, has been discovered in Cambrian beds.

Division Spermatophyta

Characteristics

Reproduce by seeds, so are commonly called seed plants.

Plants with alternation of generations (see page 145), having a very conspicuous and independent sporophyte, which is the main plant, and a very reduced gametophyte generation which is dependent on the sporophyte at all times, see page 79.

Two kinds of spores are produced, microspores and megaspores, i.e. are always heterosporous. *

There are always male and female gametophytes.

The male gametophyte develops from a microspore (pollen grain) produced inside a microsporangium (pollen sac) attached to a microsporophyll.

The female gametophyte (embryo sac) develops from a megaspore (which may also be called an embryo sac) produced inside an ovule (megasporangium) attached to a megasporophyll.

The microspores move to the megaspores in pollination.

The megaspores are retained in the ovule on the sporophyte until after pollination; after fertilisation, the ovules are shed as seeds.

Some members (gymnosperms and some dicotyledons) form true wood of secondary xylem.

Summary classification of the division

Division Spermatophyta (spermatophytes, seed plants)
  Subdivision Gymnosperms
    Class Cycadopsida (cycads)
    Class Coniferopsida (conifers)
    Class Gnetopsida
  Subdivision Angiosperms (flowering plants)
    Class Dicotyledons (Magnoliopsida, dicots)
    Class Monocotyledons (Liliopsida, monocots)

Notes

The Spermatophyta is the highest division of plants and forms the dominant vegetation today. Their structures and life cycles are well adapted to life on land. The sporophyte is usually better adapted to life on land than that of pteridophytes, and has a thicker cuticle, better xylem, and in some cases, secondary thickening.

*In spermatophytes the microspores are also called pollen grains and, especially in angiosperms, the megaspores (or particularly the female gametophytes) are called embryo sacs. This is because the life cycle of seed plants was discovered before that of pteridophytes. Later it was realised that spermatophytes evolved from pteridophyte ancestors, and that there were many homologous structures, so a second name was given to structures in spermatophytes to recognise the homology. Microsporangia are also called pollen sacs, and megasporangia are ovules (or more correctly the megasporangium is the nucellus of the ovule). Microsporophylls have no second name in gymnosperms, but are equivalent to stamens in angiosperms. Megasporophylls are called ovuliferous scales in gymnosperms and carpels in angiosperms.

Subdivision Gymnosperms

1 Flowers absent.
2 Seeds are borne naked on megasporophylls which are arranged in female cones; (the word ''gymnosperm'' means naked seed).
3 Microsporophylls are usually arranged in male cones.
4 Usually there are no sterile sporophylls (perianth) to attract insects which are not widely used in pollination; most are wind pollinated.
5 Female gametophyte has 500 or more cells with reduced archegonia.
6 Double fertilisation is absent; endosperm is made of female prothallus and is haploid.
7 One integument to ovule.
8 Most are trees or shrubs, not herbaceous plants; wood is made of tracheids only and is called softwood.
9 Phloem usually without companion cells.

Class Cycadopsida (cycads)

1 Palm-like and fern-like plants with leaves compound and frond-like.
2 Living species have motile sperm.
3 Tropical and subtropical.
Order Cycadales: only living order, includes the genera *Cycas, Dioon, Zamia.*
There are three fossil orders, of which the most primitive, the **order Pteridospermales** (pteridosperms, seed ferns) appeared in the Devonian period. These may have been ancestral to the cycads; they bear seeds on fronds which are not part of a cone. Microsporophylls are not arranged in cones.

Class Coniferopsida (conifers)

1 Not palm-like or fern-like; often tall trees with needle-like, scale-like or paddle-like leaves.
2 Sperm usually non-motile.
3 Found in cold temperate and subtropical habitats.
    Conifers are highly adapted to dry and windy habitats. To conserve water they have needle-like or scale-like leaves with sunken stomata and thick cuticles. Adaptations to windy environments are wind pollination and seed dispersal.
There are three living orders and one fossil order. The living orders are:
Order Ginkgoales
No female cones; ovules borne at tip of branches. This order includes *Ginkgo biloba*, the maidenhair tree, which is a ''living fossil''. Unlike other conifers, *Ginkgo* has motile sperm.
Order Coniferales
Female cone of many megasporophylls. This order contains most modern conifers, e.g. *Pinus* (pine), *Cupressus* (cypress), *Chamaecyparis lawsoniana* (Lawson cypress), *Juniperus* (juniper), *Cedrus* (cedar), *Abies* (fir), *Larix* (larch).
Order Taxales
No megasporophylls (or one, depending on interpretation). Ovules solitary and surrounded by an aril; e.g. *Taxus* (yew).

Class Gnetopsida

This is a group of very odd gymnosperms with only three living and very different genera. These are *Welwitschia* (the desert octopus), *Gnetum* and *Ephedra*.

Subdivision Angiosperms (flowering plants)

1 Micro- and megasporophylls are arranged in flowers.
2 Seeds are borne inside a fruit made of united megasporophylls called carpels; (the word "angiosperm" means enclosed seed).
3 Microsporophylls are modified to form stamens.
4 Sterile sporophylls (perianth) often present to attract insects for pollination; there are many mutual adaptations between insects and flowering plants; wind and insect pollination occurs.
5 Female gametophyte is very reduced, with less than 500 cells, and in many angiosperms of only 8 cells or nuclei; no archegonia are produced.
6 Double fertilisation is present; endosperm is made from the triple fusion nucleus (primary endosperm nucleus) and is triploid.
7 Two integuments to ovule.
8 In habit are trees, shrubs and herbs; wood is made of vessels as well as tracheids and in trees is called hardwood.
9 Phloem contains companion cells.

Class Dicotyledons

1 Two cotyledons in the seed.
2 Flowers have parts in fours or fives or their multiples.
3 Perianth is usually divided into petals and sepals.
4 Leaves have reticulate veins with a central midrib and lateral veins; epidermis is of irregular cells.
5 In the stem, vascular bundles are arranged in a ring around the edge of the stem.
6 In the root, there are few xylem and phloem groups.
7 Possess cambium and form true woods (hardwoods) as well as herbs and shrubs.
e.g. the families Ranunculaceae, Cruciferae, Rosaceae, Leguminosae, Labiatae, Solanaceae, Compositae.

Class Monocotyledons

1 One cotyledon in the seed.
2 Flowers have parts in threes or multiples of three.
3 Perianth usually not divided into petals and sepals.
4 Leaves have parallel veins with no central midrib; epidermis is of regular cells.
5 In the stem, vascular bundles are scattered.
6 In the root, there are many xylem and phloem groups.
7 Usually do not possess cambium and do not form true wood; palm "wood" is made of dead leaf bases; most are herbaceous.
e.g. the families Liliaceae, Amaryllidaceae, Iridaceae, Palmae, Gramineae.

Notes

In flowering plants, the orders are many and variable and are arranged in different ways by different authorities, but families are much more constant. Families with fused parts to the flower are thought to be more advanced than those with free parts.

The Compositae is considered to be the most advanced family of dicotyledons, and the Gramineae is the most advanced family of monocotyledons.

Notes on the life cycles of spermatophytes

Spermatophytes, particularly angiosperms, have many adaptations that are improvements on pteridophytes, which is why they are the dominant and most successful vegetation on land today.

1 The sporophyte is the main plant, the cycad, conifer or flowering plant. It is usually better adapted to life on land than pteridophytes, with a thicker cuticle, better xylem and in some cases, secondary thickening.

2 The gametophyte is not free living, so there is no vulnerable haploid phase. It is always protected on the sporophyte.

3 The male gamete does not swim to the ovum in free external water. It is taken to the female gamete by the growth of the pollen tube, so free water is not necessary for normal sexual reproduction, and the plant can live in a much drier habitat.
   The lack of need for free water also means that the plant can grow taller, and does not need a low growing phase (since motile sperm cannot swim up heights). Cycads are interesting in that they do have motile sperm (see below).

4 The reproductive structure is a seed not a spore. This is an advantage because:
(a) It contains a food store provided by the parent sporophyte.
(b) It contains a multicellular embryo which has started its development while on the sporophyte, and will take less time to become established in a habitat; it can compete with larger plants better than a structure developing from a spore.

5 In angiosperms, animals are used in the life cycle as well as wind:
(a) Insects as well as wind take pollen from flower to flower in pollination.
(b) Birds and mammals are used as well as wind in fruit dispersal.
Animals are not "better" than wind, but using animals enables the plants to live in wind-free habitats, and so to exploit a greater range of ecological niches.

The significance of the life cycle of cycads

Cycads are unusual in that they have motile male gametes with flagella. This is not "necessary" in the way that it is in pteridophytes where the spermatozoids have to swim in water on the ground to reach the ovum. In cycads, pollination occurs and there is a pollen tube which grows to the ovum, but then the male gamete swims in liquid in the ovule to reach the ovum. It is suggested that swimming sperm is a vestigial characteristic, indicating that cycads evolved from pteridophyte ancestors.

Cycas species cycad	*Division*	Spermatophyta
	Subdivision	Gymnosperms
	Class	Cycadopsida
	Order	Cycadales
	Genus	*Cycas*

Entire mature female sporophyte: parts above ground shown

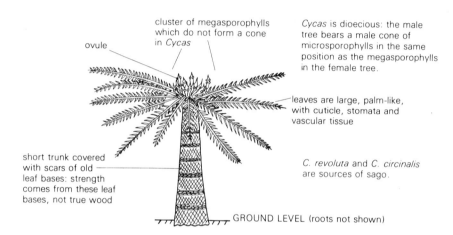

ovule

cluster of megasporophylls which do not form a cone in *Cycas*

Cycas is dioecious: the male tree bears a male cone of microsporophylls in the same position as the megasporophylls in the female tree.

leaves are large, palm-like, with cuticle, stomata and vascular tissue

short trunk covered with scars of old leaf bases: strength comes from these leaf bases, not true wood

C. revoluta and *C. circinalis* are sources of sago.

GROUND LEVEL (roots not shown)

Male cone of _Cycas_

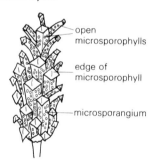

open microsporophylls

edge of microsporophyll

microsporangium

Female cone of _Dioon_, partly dissected

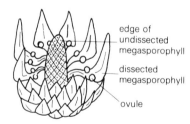

edge of undissected megasporophyll

dissected megasporophyll

ovule

Dioon has 2 ovules per megasporophyll, and its megasporophylls are arranged in cones.

Species of *Cycas* have between 2 and 8 ovules per megasporophyll.

Distribution: *Cycas* extends from Japan to Queensland, and is also found in India and Madagascar. Other living genera of cycads are native to Central America, South Africa, eastern Asia and Australia. In the past they had an almost world-wide distribution and made important contributions to Mesozoic forests.

Notes: most cycads have both microsporophylls and megasporophylls arranged in male and female cones respectively. *Cycas* is unusual in that, although its microsporophylls are in male cones, its megasporophylls are borne in a loose cluster at the tip of the trunk, which is thought to be a primitive feature.

Primitive forms of Cycadopsida, the pteridosperms (seed ferns), appeared in late Devonian times, and the Cycadales (true cycads) first appeared in the Triassic period; they are now regarded as "living fossils".

Life cycle of a cycad

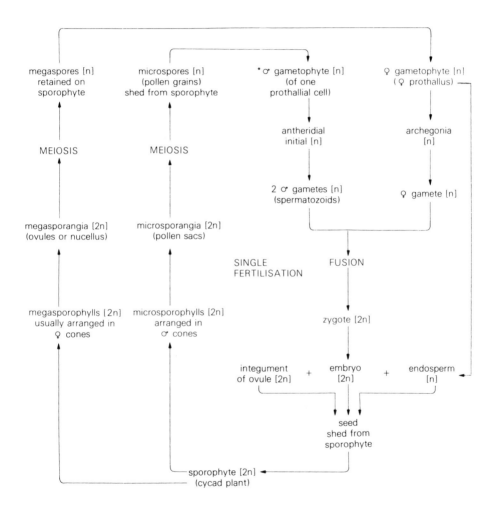

*In cycads, the male gametophyte is formed as follows:

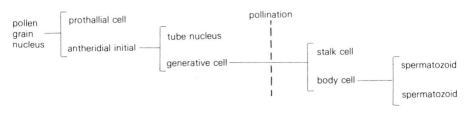

Cycas species	Division	Spermatophyta
cycad	Subdivision	Gymnosperms
	Class	Cycadopsida
life cycle	Order	Cycadales
male structures	Genus	Cycas

One microsporophyll: lower surface   **V.S. part of one microsporophyll**

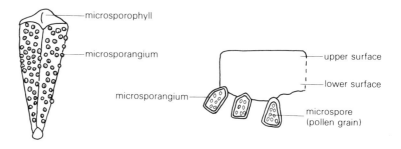

L.S. pollen grains during their development

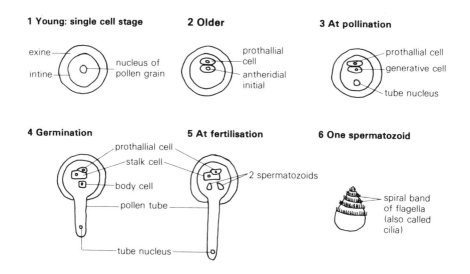

For explanation, see page 84.

Notes: cycads are completely adapted to life on land and do not need the external water for fertilisation by a free swimming male gamete that is required by bryophytes and pteridophytes. But cycads have motile sperm with cilia, which swims to the archegonium in liquid within the ovule, once the pollen tube has grown to the archegonium. There is no ''need'' for swimming sperm, and it is thought to be a primitive characteristic, suggesting that cycads evolved from pteridophyte ancestors.

Division	Spermatophyta
Subdivision	Gymnosperms
Class	Cycadopsida
Order	Cycadales
Genus	Cycas

Cycas species
cycad

life cycle
female structures

One megasporophyll (ovuliferous scale)

The number of ovules may vary from 2 to 8 depending on species.

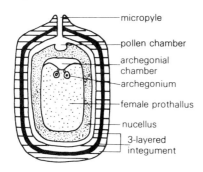

- ovule
- megasporophyll

L.S. young ovule

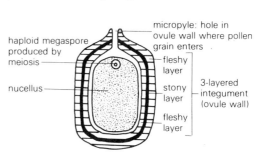

- haploid megaspore produced by meiosis
- nucellus
- micropyle: hole in ovule wall where pollen grain enters
- fleshy layer
- stony layer
- fleshy layer
- 3-layered integument (ovule wall)

L.S. ovule just before fertilisation

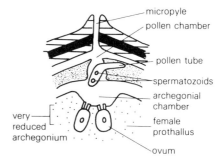

- micropyle
- pollen chamber
- archegonial chamber
- archegonium
- female prothallus
- nucellus
- 3-layered integument

Enlarged micropyle end of ovule to show fertilisation

- micropyle
- pollen chamber
- pollen tube
- spermatozoids
- archegonial chamber
- very reduced archegonium
- female prothallus
- ovum

L.S. seed

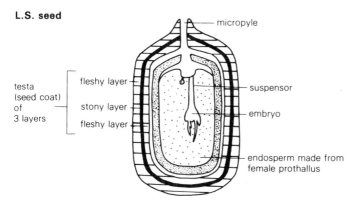

- micropyle
- testa (seed coat) of 3 layers
  - fleshy layer
  - stony layer
  - fleshy layer
- suspensor
- embryo
- endosperm made from female prothallus

For explanation, see page 84.

Life cycle of a cycad: *Cycas*

Cycads show alternation of generations in which the sporophyte is the most conspicuous generation, forming the cycad plant, and is the independent generation, well adapted to life on land. The gametophyte is very reduced and is always dependent on the sporophyte. As in all spermatophytes, cycads are heterosporous. The concept of alternation of generations is explained on page 145, and diagrams of cycad structure and life cycle are shown on pages 80 to 83. The sporophyte produces haploid microspores (pollen grains) and megaspores by meiosis in micro- and megasporangia borne on micro- and megasporophylls usually arranged in male and female cones.

The male structures consist of woody microsporophylls organised into male cones. Each microsporophyll has many microsporangia (pollen sacs) on its lower surface. Meiosis occurs in the pollen sacs to produce microspores (pollen grains). At first each pollen grain is a single-celled structure with two coats, an exine and intine. The single cell divides to form a prothallial cell and an antheridial initial, which divides to form a tube nucleus and generative cell, so there are now three cells in the pollen grain. Pollination occurs at this stage: the pollen sacs dehisce and pollen is blown in the wind to the megasporophyll and lands on the micropyle of the ovule. It is caught in the pollination drop, a drop of liquid secreted by the ovule from the micropyle.

The female structures consist of woody megasporophylls usually organised into female cones (but not in *Cycas* where they are loosely arranged). Each megasporophyll has a number of megasporangia (ovules) attached at the side. Each ovule has a coat (integument) of three layers, an outer and inner fleshy layer and a middle stony layer. Meiosis occurs in the ovule in a region called the nucellus, to produce four haploid cells. Three of these abort, and the remaining one forms the megaspore which is retained on the sporophyte.

In the megaspore, the single cell divides to form the female gametophyte which consists of female prothallus and archegonia and crushes the nucellus. The female prothallus grows up on either side of the archegonia so that they lie in a depression, the archegonial chamber. The nucellus grows upwards towards the micropyle and a cavity develops in it, known as the pollen chamber. Pollen grains caught in the pollination drop are drawn into the pollen chamber. Here a short pollen tube is produced which grows laterally along the nucellus; above the archegonia it secretes enzymes which digest material away above the archegonia, so that the archegonial and pollen chambers are united.

Meanwhile the generative cell divides into a stalk cell and a body cell, which in turn divides into two flagellated spermatozoids. The spermatozoids of cycads are the largest known, up to 0.3 mm in length. The pollen tube ruptures, liberating fluid into the pollen chamber. The spermatozoids swim in this fluid to the archegonia. The fluid has a high osmotic pressure, and when it comes into contact with the necks of the archegonia, they lose water by osmosis and become flaccid, so that the spermatozoids have access to the ova. One swims down the neck and fuses with the ovum to form a zygote.

After fertilisation, the ovule develops into a seed. The integument forms the seed coat (testa) consisting, like the integument, of three layers. The zygote develops into the embryo and the female prothallus forms the food store (endosperm). The seeds are animal-dispersed, and the stony layer protects the delicate embryo from digestion. Once dispersed, the seeds grow into new sporophytes.

Division	Spermatophyta
Subdivision	Gymnosperms
Class	Coniferopsida
Order	Ginkgoales
Genus	Ginkgo

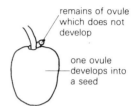

Ginkgo biloba
maidenhair tree

Short shoot with ovules

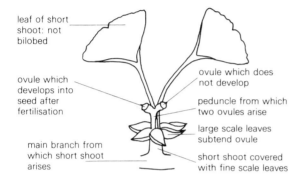

leaf of short shoot: not bilobed

ovule which develops into seed after fertilisation

main branch from which short shoot arises

ovule which does not develop

peduncle from which two ovules arise

large scale leaves subtend ovule

short shoot covered with fine scale leaves

Ripe seed

remains of ovule which does not develop

one ovule develops into a seed

Outer layer of testa is yellow and fleshy and gives off a smell of rancid butter; middle layer is stony, and inner layer is watery.

Short shoot with microsporangia

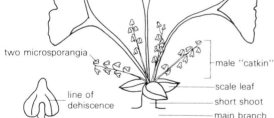

two microsporangia

line of dehiscence

male "catkin"

scale leaf

short shoot

main branch

Two microsporangia

Short shoots are special shoots bearing reproductive structures.

Bilobed leaf from long shoot

two lobes

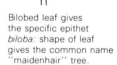

Bilobed leaf gives the specific epithet *biloba:* shape of leaf gives the common name "maidenhair" tree.

Long shoots bear normal foliage leaves.

Distribution: only known to grow wild in inland China, but the tree was widely cultivated in temple gardens of China and Japan, from where it was brought to Europe in about 1727. It is now widely planted as an ornamental, usually as the male, because the female produces a foul-smelling seed.

Notes: *Ginkgo biloba* is a "living fossil"; most of its relatives became extinct at the end of the Cretaceous period, but fossils identical to *Ginkgo* are known from the Triassic period and the order goes back to Carboniferous times.

The tree is dioecious, and sex is determined by a pair of sex chromosomes, XX in the female and XY in the male. The seed is eaten roasted in China and Japan; the edible part is female prothallus.

85

Pinus sylvestris	Division	Spermatophyta
Scots pine	Subdivision	Gymnosperms
	Class	Coniferopsida
	Order	Coniferales
	Genus	Pinus

Branch of Scots pine

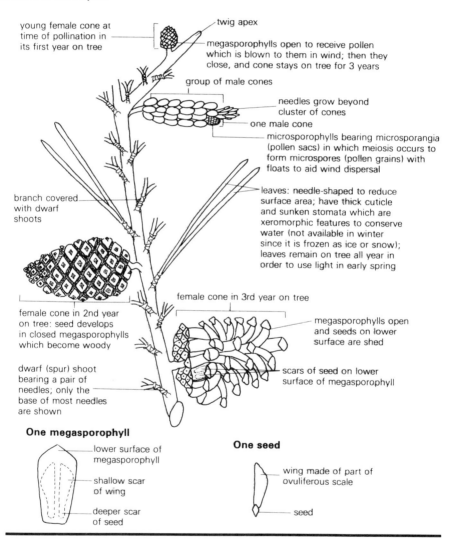

young female cone at time of pollination in its first year on tree

twig apex

megasporophylls open to receive pollen which is blown to them in wind; then they close, and cone stays on tree for 3 years

group of male cones

needles grow beyond cluster of cones

one male cone

microsporophylls bearing microsporangia (pollen sacs) in which meiosis occurs to form microspores (pollen grains) with floats to aid wind dispersal

branch covered with dwarf shoots

leaves: needle-shaped to reduce surface area; have thick cuticle and sunken stomata which are xeromorphic features to conserve water (not available in winter since it is frozen as ice or snow); leaves remain on tree all year in order to use light in early spring

female cone in 3rd year on tree

female cone in 2nd year on tree: seed develops in closed megasporophylls which become woody

megasporophylls open and seeds on lower surface are shed

dwarf (spur) shoot bearing a pair of needles; only the base of most needles are shown

scars of seed on lower surface of megasporophyll

One megasporophyll

lower surface of megasporophyll

shallow scar of wing

deeper scar of seed

One seed

wing made of part of ovuliferous scale

seed

Distribution: *Pinus sylvestris* is native to most of Europe and N. Asia, particularly on mountains and on well drained soils. It is adapted to cold, windy conditions by its xeromorphic leaves, and use of wind for pollination and seed dispersal.
Notes: some species of *Pinus* are found in subtropical climates with a dry season, since the needle-like leaves can conserve water in drought.

Life cycle of a conifer

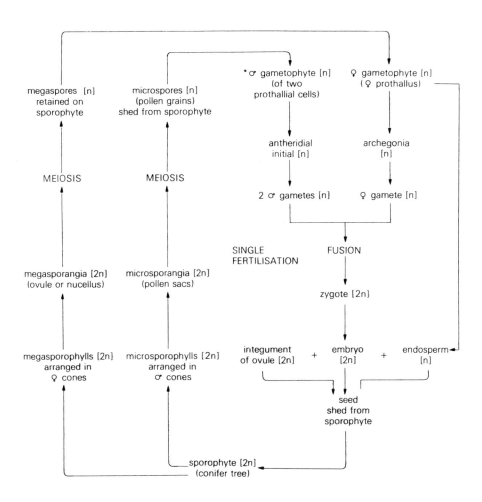

* In many conifers, the male gametophyte is formed as follows:

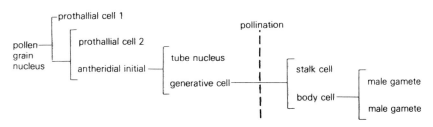

Pinus species pine **life cycle** **male structures**	Division	Spermatophyta
	Subdivision	Gymnosperms
	Class	Coniferopsida
	Order	Coniferales
	Genus	_Pinus_

L.S. male cone

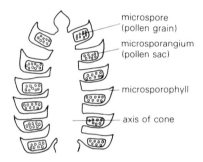

microspore (pollen grain)

microsporangium (pollen sac)

microsporophyll

axis of cone

One microsporophyll, lower surface

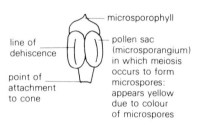

microsporophyll

line of dehiscence

point of attachment to cone

pollen sac (microsporangium) in which meiosis occurs to form microspores: appears yellow due to colour of microspores

L.S. stages of development of pollen grain (microspore)

1 Before shedding

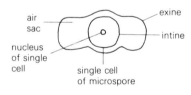

air sac

nucleus of single cell

exine

intine

single cell of microspore

2 At time of pollination

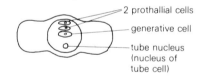

2 prothallial cells

generative cell

tube nucleus (nucleus of tube cell)

3 Germinating pollen grain

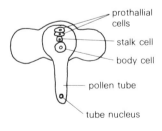

prothallial cells

stalk cell

body cell

pollen tube

tube nucleus

4 Pollen grain just before fertilisation

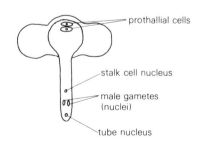

prothallial cells

stalk cell nucleus

male gametes (nuclei)

tube nucleus

For explanation, see page 90.

Division	Spermatophyta	
Subdivision	Gymnosperms	
Class	Coniferopsida	
Order	Coniferales	
Genus	*Pinus*	

Pinus species
pine

**life cycle
female structures**

L.S. female cone

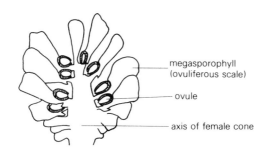

- megasporophyll (ovuliferous scale)
- ovule
- axis of female cone

L.S. young ovule on megasporophyll

nucellus

integument

ovuliferous scale

megaspore

bract scale

L.S. mature ovule

integument   nucellus

micropyle

ovule

part of ovuliferous scale

female prothallus   ovum in archegonium

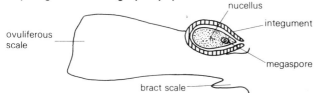

One archegonium

neck

ovum in venter

female prothallus

L.S. seed

testa

food store (endosperm) made of female prothallus

seed

micropyle

wing

part of ovuliferous scale

nucellus is squashed out of existence

cotyledons   plumule   radicle

embryo

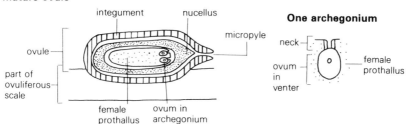
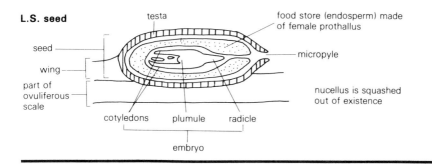

For explanation, see page 90.

Life cycle of a conifer: *Pinus*

Conifers show alternation of generations in which the sporophyte is the most conspicuous generation, forming the conifer tree, and is the independent generation well adapted to life on land. The gametophyte is very reduced and is always dependent on the sporophyte. As in all spermatophytes, conifers are heterosporous. The concept of alternation of generations is explained on page 145, and diagrams of conifer structure and life cycle are shown on pages 86 to 89. The sporophyte produces haploid microspores (pollen grains) and megaspores (sometimes called embryo sacs) by meiosis in micro- and megasporangia borne on micro- and megasporophylls arranged in male and female cones.

The male structures are male cones made up of scaly microsporophylls on the lower surface of which are two microsporangia (pollen sacs). Meiosis occurs in the pollen sacs to produce microspores (pollen grains). The pollen grain wall is made of two layers, an intine and an exine which is expanded into air floats so that the grain can float in air currents. Inside the pollen grain the single haploid nucleus divides three times to form the male gametophyte. This consists of two prothallial cells and a generative cell embedded in a large tube cell. At this stage the pollen sacs dehisce, and pollen grains are liberated into the air and dispersed by wind. Some are blown to the female cone whose scales open to receive them. The pollen is caught in the pollination drop, a drop of liquid secreted by the ovule.

The female structures are female cones, which are borne singly and take three years to mature. Each cone consists of woody megasporophylls, called ovuliferous scales in conifers, with a small bract scale attached to each. Two megasporangia (ovules) are borne side by side on the lower surface of each scale. Each ovule is surrounded by a single integument with a micropyle. A cell in the nucellus of the ovule undergoes meiosis to produce four haploid cells, three of which abort, while the remaining one becomes the megaspore. This undergoes haploid mitosis to give the female gametophyte which consists of female prothallus with one or two archegonia near the micropyle. At some point during the development of the female gametophyte the ovuliferous scales open and pollination occurs. The scales then close up again and further development proceeds.

When the archegonia are fully formed the ovule is ready for fertilisation. The pollen grain produces a pollen tube which grows through the micropyle to the archegonium, with the nucleus of the tube cell at the tip. The generative cell divides into a stalk cell and a body cell. The nucleus of the body cell divides into two; these two nuclei act as male gametes and join the tube nucleus at the tip of the pollen tube. One male gamete passes down the neck of the archegonium and fuses with the ovum to form a zygote.

The ovule now develops into a seed. The zygote divides to form an embryo consisting of a radicle, plumule and many cotyledons. The integument forms the testa (seed coat). The endosperm (food store) is made of female prothallus and is haploid. Since each ovule forms a seed, two seeds are attached to the lower surface of each ovuliferous scale. In *Pinus* the seed has a wing which is made from the surface layers of the scale, but many conifers do not have winged seeds. In dry weather the ovuliferous scales move apart and the seeds are shed and dispersed by wind.

		Division	Spermatophyta		
		Subdivision	Gymnosperms		
		Class	Coniferopsida		
Order	Coniferales			Order	Taxales
Genus	Chamaecyparis			Genus	Taxus

Lawson cypress: *Chamaecyparis lawsoniana*

Branch with young female cone

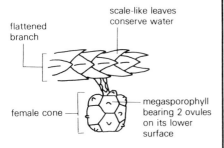

flattened branch

scale-like leaves conserve water

female cone

megasporophyll bearing 2 ovules on its lower surface

Megasporophylls separate to release seeds when cone becomes older.

Branch with young male cone

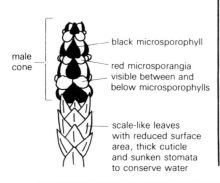

male cone

black microsporophyll

red microsporangia visible between and below microsporophylls

scale-like leaves with reduced surface area, thick cuticle and sunken stomata to conserve water

Yew: *Taxus baccata*

Branch with old female cone

seed with hard black testa to resist digestion

red fleshy aril, eaten by birds, ensuring seed dispersal

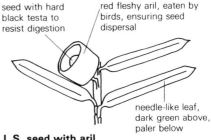

needle-like leaf, dark green above, paler below

L.S. seed with aril

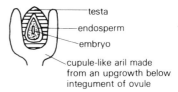

testa

endosperm

embryo

cupule-like aril made from an upgrowth below integument of ovule

Branch with young male cone

one microsporophyll with usually about 6 lobes, bearing microsporangia on lower surface

bud scale

male cone

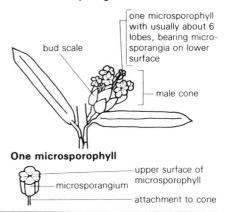

One microsporophyll

upper surface of microsporophyll

microsporangium

attachment to cone

Distribution: Lawson cypress is widely planted as an ornamental, but is native to north-western U.S.A. Other cypresses are also native to southern Europe and Asia.
Notes: male and female cones are borne on the same tree. Cypresses have scale-like leaves which conserve water in a similar way, and for the same reasons, as conifer needles.

Distribution: yew is widely distributed in Europe, Asia and North America, and is often planted as an ornamental.
Notes: there are separate male and female trees. The male cone consists of microsporophylls, but the female ''cone'' is a single ovule, which when mature, becomes surrounded by a red fleshy aril, or cupule, made from the integument.

Life cycle of an angiosperm: generalised flowering plant

Angiosperms show alternation of generations in which the sporophyte is the most conspicuous generation, forming the flowering plant, and is the independent generation well adapted to life on land. The gametophyte is very reduced and is always dependent on the sporophyte. The concept of alternation of generations is explained on page 145, and diagrams concerning the life cycle of a flowering plant are on pages 94 to 99.

The sporophyte produces microspores (pollen grains) and megaspores (sometimes called embryo sacs) by meiosis in a special organ, the flower, whose structure is shown on page 94. Since the gametophyte is so reduced, and fusion of gametes occurs in the flower, the flower is also the organ of sexual reproduction. The outer layers of the flower are made of sterile sporophylls. Sometimes they are all the same and are called perianth lobes or tepals, or sometimes, as in the diagram, they are divided into outer sepals and inner petals.

The male parts of the flower are the stamens (microsporophylls), each consisting of an anther and filament. Inside the anther are four compartments called pollen sacs (microsporangia) in which meiosis occurs in pollen mother cells to produce pollen grains (microspores). The pollen grains stick together when young to form a tetrad of four, which are the products of one meiosis. Male structures are shown on page 96.

The female parts of the flower are the carpels (megasporophylls) that make up the stigma, style and ovary, and in which ovules arise. The ovules contain a uniform mass of cells, the nucellus. Meiosis will occur in the nucellus in megaspore mother cells to produce the megaspore, which may be considered a unicellular embryo sac. The nucellus, rather than the whole ovule, is truly equivalent to the megasporangium, since integuments are found only in spermatophytes and have no equivalent in pteridophytes. Female structures are shown on page 97.

The pollen grains are shed when the anther dehisces, which happens in dry conditions. The two pollen sacs on each side fuse to form a single one at a thin part of the wall called the stomium; then the wall breaks there and curls back due to differential drying out of the fibrous layer in the wall. Pollen travels from the open anther to the stigma, usually of another flower. It may be carried by wind or insects in pollination, see page 100. Pollination in individual flowers is described with the flower diagrams.

When the pollen grain is shed it is single-celled, with a coat of two layers, an intine and a very resistant sculptured exine. When it lands on the stigma, it begins to develop. The single haploid nucleus divides into two by mitosis to form a tube nucleus and generative nucleus. The generative nucleus divides into two male gametes. The intine pushes out to form a pollen tube, and the gametes migrate into the tube which grows down to the ovule, through the style. There is no prothallus or antheridium. The male gametophyte consists only of a tube nucleus and two male gametes.

Meanwhile the female part is developing. A cell arises in the nucellus, called the megaspore mother cell or embryo sac mother cell, which undergoes meiosis to form a line of haploid cells. Three of these abort and the remaining one forms the megaspore (or unicellular embryo sac). The female gametophyte, which is called the embryo sac, arises in this, in the ovule, and is not shed from the sporophyte. The haploid nucleus divides three times to form eight nuclei, four at each end of the embryo sac. Two of these nuclei migrate back to the centre of the embryo sac and are known as polar nuclei; they may fuse to form a diploid nucleus called the secondary or definitive nucleus. At this stage the female gametophyte, or embryo sac, consists of seven or eight nuclei. The centre of the three nuclei near the micropyle is the ovum; the cells on either side of it are called synergids, and those at the other end of the embryo sac are called antipodal cells. These structures may be called cells because each nucleus has some cytoplasm and a membrane around it, but cell walls are not formed. The situation of all stages is shown on page 98.

The pollen tube grows down the style, feeding on sugars from the stigma, and either digesting the style cells as it grows, or pushing its way between cells. It is probably growing away from oxygen in air (negatively aerotropic) or possibly towards chemicals in the ovary. It enters the ovule via the micropyle, probably guided at the end by chemicals secreted by the embryo sac. Once at the ovule, the tube nucleus disintegrates leaving the male gametes at the end of the pollen tube.

Fertilisation now occurs, and each of the two male gametes undergoes a fusion:
1 One male gamete fuses with the ovum to form a zygote.
2 The other male gamete fuses with the two haploid polar nuclei to give a triploid nucleus called the triple fusion nucleus or primary endosperm nucleus. If the secondary nucleus has formed, the male gamete fuses with this to give a triploid nucleus.
Since two male gametes fuse, this is called double fertilisation.

The diploid zygote undergoes mitosis to form an embryo (i.e. an embryonic sporophyte) consisting of a radicle (embryonic root), a plumule (embryonic shoot) and one cotyledon (in monocots) or two cotyledons (in dicots). The embryo is attached to the wall of the embryo sac by a suspensor, through which it is sometimes nourished.

The triple fusion nucleus undergoes mitosis to form the endosperm (food store). This may or may not be absorbed into the cotyledons. If it is, the seed is non-endospermic. If it is not absorbed, the seed is endospermic. Thin cell walls are usually formed, and the endosperm becomes a semi-fluid food storage tissue, the cells of which become filled with carbohydrates and/or lipids. The nucellus becomes crushed out of existence as the embryo and endosperm grow. It is thought that the double fertilisation ensures that endosperm is only formed in fertilised ovules.

The above processes result in the production of hormones, especially auxins, which cause further changes:
1 The integuments become the testa (seed coat) by becoming hard and dry and impregnated with lignin and cutin. Technically the outer integument becomes the testa and the inner becomes the tegmen. Both are tough and protective.
2 The ovary wall becomes the pericarp (fruit wall) by becoming succulent or dry.
3 The ovary becomes the fruit.
4 The ovules become seeds by losing water and drying out.
5 The stigma, style and perianth usually wither, but various parts of the flower may remain attached to the fruit to aid dispersal.
The situation is now as shown on page 99.

The seed, or in some cases the whole fruit is shed from the sporophyte and usually dispersed some distance, using wind, animals, or self dispersive mechanisms, often an explosive device. The seed often undergoes a period of dormancy and, when conditions are suitable, it germinates to form a new sporophyte.

Half flower of a generalised flower

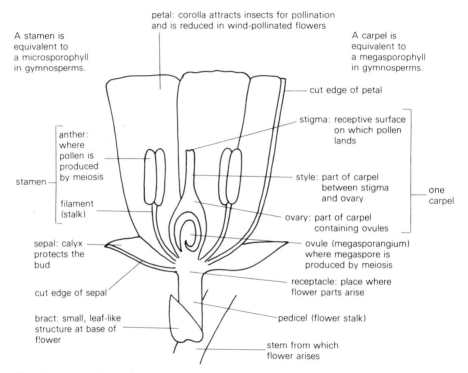

petal: corolla attracts insects for pollination and is reduced in wind-pollinated flowers

A stamen is equivalent to a microsporophyll in gymnosperms.

A carpel is equivalent to a megasporophyll in gymnosperms.

cut edge of petal

stigma: receptive surface on which pollen lands

anther: where pollen is produced by meiosis

stamen

filament (stalk)

style: part of carpel between stigma and ovary

ovary: part of carpel containing ovules

one carpel

sepal: calyx protects the bud

ovule (megasporangium) where megaspore is produced by meiosis

cut edge of sepal

receptacle: place where flower parts arise

bract: small, leaf-like structure at base of flower

pedicel (flower stalk)

stem from which flower arises

The calyx and corolla may be considered as sterile sporophylls.
Nectar may be produced in nectaries at the base of the petals,
in the ovary wall, receptacle, or base of the stamens.

The sepals are collectively called the calyx.
The petals are collectively called the corolla.
The calyx and corolla are collectively called the perianth.
The stamens are collectively called the androecium.
The carpels are collectively called the gynaecium.

The above flower is hermaphrodite, but flowers may also be unisexual, either male (staminate) in which the carpels are missing, or female (carpellate) in which the stamens are missing. Here, all parts of the flower are free, but more advanced flowers have varying degrees of fusion of parts. In this example the flower is radially symmetrical (actinomorphic) but flowers may also be bilaterally symmetrical (zygomorphic). Fusion of parts and zygomorphy are thought to be advanced features, and adaptations to specialised insect pollination. Here, the ovary contains one ovule, but there may be many ovules per carpel and/or many carpels per ovary.

Life cycle of a flowering plant

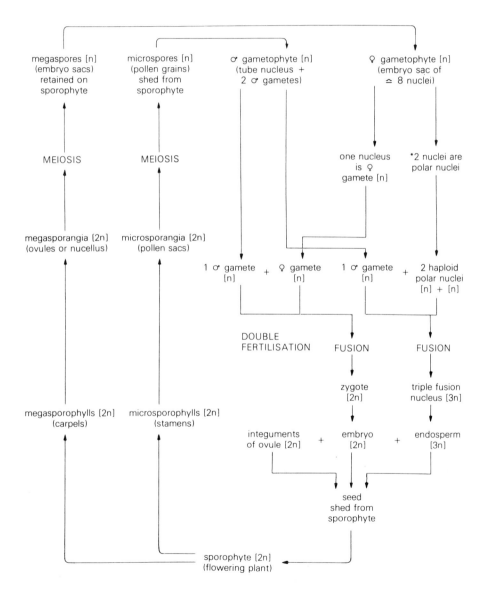

*These two nuclei may fuse together. The diploid nucleus formed is then called the secondary nucleus and fuses with a male gamete to form the triploid triple fusion nucleus, which is also called the primary endosperm nucleus.

Structure of anther

young anther: entire **top of anther, cut off across A–B** **T.S. across A–B**

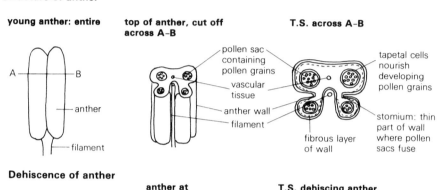

A ———— B

anther

filament

pollen sac containing pollen grains

vascular tissue

anther wall

filament

tapetal cells nourish developing pollen grains

fibrous layer of wall

stomium: thin part of wall where pollen sacs fuse

Dehiscence of anther

anther at dehiscence **T.S. dehiscing anther**

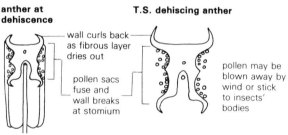

wall curls back as fibrous layer dries out

pollen sacs fuse and wall breaks at stomium

pollen may be blown away by wind or stick to insects' bodies

Development of pollen grain (L.S.)

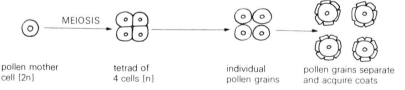

MEIOSIS

pollen mother cell [2n]

tetrad of 4 cells [n]

individual pollen grains

pollen grains separate and acquire coats

Development of male gametophyte in pollen grain (L.S.)

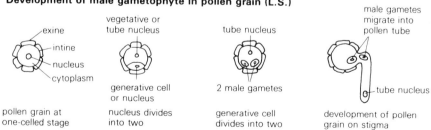

exine

intine

nucleus

cytoplasm

vegetative or tube nucleus

generative cell or nucleus

tube nucleus

2 male gametes

male gametes migrate into pollen tube

tube nucleus

pollen grain at one-celled stage

nucleus divides into two

generative cell divides into two

development of pollen grain on stigma

For explanation, see page 92.

Division	Spermatophyta
Subdivision	Angiosperms

Development of ovule

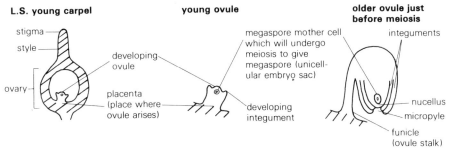

L.S. young carpel

stigma
style
ovary
developing ovule
placenta (place where ovule arises)

young ovule

megaspore mother cell which will undergo meiosis to give megaspore (unicellular embryo sac)
developing integument

older ovule just before meiosis

integuments
nucellus
micropyle
funicle (ovule stalk)

most ovules bend over as they develop (anatropous)

Development of female gametophyte in ovule

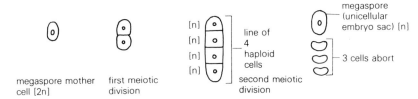

megaspore mother cell [2n]

first meiotic division

[n]
[n]
[n]
[n]
line of 4 haploid cells

second meiotic division

megaspore (unicellular embryo sac) [n]

3 cells abort

Haploid mitosis in megaspore to form female gametophyte (embryo sac)

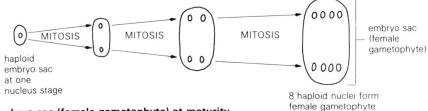

haploid embryo sac at one nucleus stage

MITOSIS   MITOSIS   MITOSIS

embryo sac (female gametophyte)

8 haploid nuclei form female gametophyte

Embryo sac (female gametophyte) at maturity

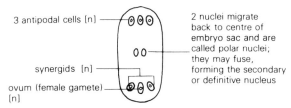

3 antipodal cells [n]

synergids [n]

ovum (female gamete) [n]

2 nuclei migrate back to centre of embryo sac and are called polar nuclei; they may fuse, forming the secondary or definitive nucleus

At maturity, the embryo sac fills the ovule.

For explanation, see page 92.

L.S. one carpel just before fertilisation

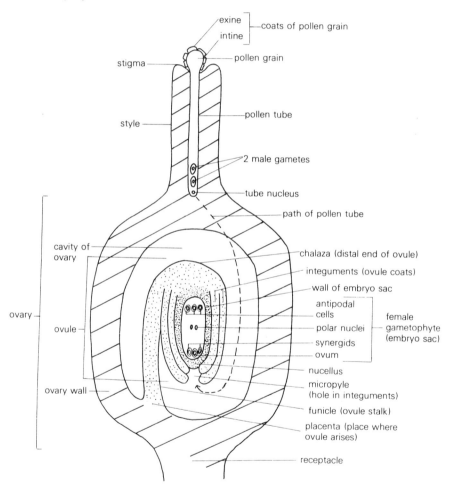

One male gamete fuses with the female gamete to form a zygote which develops into the embryo.

The other male gamete fuses with the polar nuclei to form a triploid nucleus which develops into the endosperm.

The two polar nuclei may fuse to form a secondary nucleus which then fuses with the male gamete to form the triploid nucleus which develops into the endosperm.

For explanation, see page 92.

Division	Spermatophyta	
Subdivision	Angiosperms	

Generalised flower

life cycle
fruit and seed
formation

L.S. one carpel after fertilisation

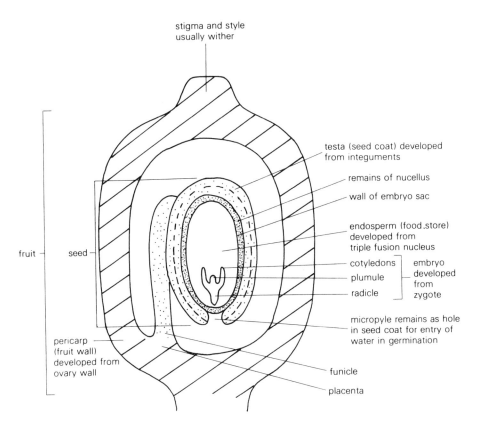

stigma and style
usually wither

testa (seed coat) developed
from integuments

remains of nucellus

wall of embryo sac

endosperm (food.store)
developed from
triple fusion nucleus

cotyledons ⎫ embryo
plumule ⎬ developed from
radicle ⎭ zygote

micropyle remains as hole
in seed coat for entry of
water in germination

fruit

seed

pericarp
(fruit wall)
developed from
ovary wall

funicle

placenta

The fruit is developed from the ovary.
The seed is developed from the ovule.

For explanation, see page 92.
The seed shown here is a dicotyledon since it has two cotyledons.
It is an endospermic seed, as the cotyledons do not absorb endosperm.
The pericarp may become fleshy (succulent fruit) for animal dispersal, or may
become dry and hard (dry fruit). Although the stigma, style, and other floral parts
usually wither, they may remain attached to the fruit and aid dispersal, e.g.
buttercup (page 104).

Pollination

Pollination is the transfer of pollen from an anther to a stigma.
Self pollination is pollination within the same flower, and causes inbreeding.
Cross pollination is pollination in which pollen is carried to the stigma of a different flower, usually of a different plant, and encourages outbreeding.

Most flowers try to encourage some cross pollination, since outbreeding produces variation necessary to the evolution of the species. Many plants have mechanisms which allow selfing to occur if crossing fails.

Cross pollination may be caused or encouraged in the following ways:

1 Dioecious plants: there are separate male and female plants, so pollination must occur between two plants, e.g. hop, yew, willow.

2 Monoecious plants: there are separate male and female flowers on the same plant, so pollen has to travel from one flower to another. If the flowers mature at different times, this encourages cross pollination, e.g. maize.

Most plants have hermaphrodite flowers, and use one of the following techniques:

3 Dichogamous or heterogamous flowers: the anthers and stigmas mature at different times. If the anthers mature first, the flower is said to be **protandrous;** pollen cannot grow on its own immature stigma and will be carried to a more mature plant. If the stigmas mature first, the flower is said to be **protogynous;** pollen from another plant will probably reach the stigma before its own pollen is ripe. There is usually an overlap period to permit selfing if crossing has failed. If the anthers and stigmas mature at the same time, the flower is said to be **homogamous.**

4 Incompatibility: pollen will not grow on its own stigma, or that of a plant of identical genotype, whether essential organs mature at the same time or at different times. Incompatibility is genetically controlled and is found, for example, in primroses, cherries and apples.
   The methods of obtaining incompatibility vary. In some cases, proteins produced on the stigma and pollen grain coat interact so that the grain does not grow. In other cases, foreign pollen grows faster than the plant's own pollen, or may produce longer pollen tubes, because plugs of callose, a polysaccharide, are produced by the style below pollen grains of identical genotype.
   A few plants have an externally visible condition of incompatibility called **heterostyly,** where flowers exist with two or sometimes three lengths of style. The classic example is the primrose where there are two types of flower, pin-eyed with a long style, and thrum-eyed with a short one. Stamens in a pin-eyed flower are at the same height as the stigma in a thrum-eyed flower, and vice versa. Pollination can only occur from a stamen to a stigma of a corresponding height.

Vectors used in pollination

The agent carrying pollen is called a vector. In temperate climates the wind and insects are the most usual vectors, but water is used occasionally. In the tropics bats and birds are quite common pollinators, when visiting flowers for nectar, e.g. humming birds which hover in front of flowers to suck nectar.
   Having an unspecialised flower with free parts is thought to be primitive, since it can be pollinated by many insects. Fusion of parts is considered to be an advanced feature, since it leads to specialisation so that only a few insects can reach the nectar or pollen.

Wind pollination: anemophily

Characteristics of wind pollinated (anemophilous) flowers

1 Flowers small or numerous, inconspicuous, with small, often green petals, and no scent or nectar.
2 Anthers large and often dangling out of the flower. Inflorescence often hangs loosely so that the smallest movement of wind liberates pollen.
3 Pollen light, smooth, produced in large quantities, since much is wasted.
4 Stigmas are often feathery and hang outside flower, to provide a large surface area to catch pollen.

Insect pollination: entomophily

Characteristics of insect pollinated (entomophilous) flowers

1 Flowers usually large, or grouped into large inflorescences. Petals brightly coloured and often scented to attract insects. Nectar secreted by nectaries, or large amounts of pollen produced for insect food; petals may be marked with honey guides to direct insects to nectar.
2 Anthers smaller and usually inside flower in fixed position where insects are likely to brush against them.
3 Pollen is often sticky or spiky to attach to insects; less pollen is produced than in wind pollinated flowers as less is wasted.
4 Stigma is flat, lobed or forked, and in a fixed position inside the flower where pollen-collecting insects are likely to brush it.

Such flowers attract a great range of insects, including flies, beetles, bees, butterflies and moths. Some flowers are particularly adapted to butterflies or bees, with nectar that can only be reached with a long tongue (butterflies) or by the insect forcing its way into the flower (bees).

There are flowers with very specialised insect pollination. Carrion flowers look and smell like rotting meat to attract flies, e.g. cuckoo pint and *Rafflesia*. Some flowers, such as cuckoo pint and lady's slipper orchid, trap insects in the base of the inflorescence or flower and hold them overnight to cause pollination. Orchids called pseudocopulatory orchids resemble a female insect and encourage a male to visit the flower to try to mate with it, and in doing so it picks up or deposits pollen. Very extreme adaptations occur where an insect lays its eggs in a flower or inflorescence. This is seen in the *Yucca* flower which is pollinated by the yucca moth, and in the fig inflorescence which is pollinated by the fig wasp.

Water pollination: hydrophily

This is quite rare as most water plants have flowers above water which are pollinated by insects or wind, but it is found in the pondweeds *Elodea* and *Valisneria* and in the marine grass, *Zostera*.

Bird and bat pollination

These vectors are found mainly in the tropics and subtropics.

Bird flowers are similar to insect flowers but with much nectar and usually no scent, e.g. *Banksia* of Australia.

Bat flowers open at night and have a sour musty scent, dingy colour, and are large enough to support a bat's weight, e.g. baobab, durian and kapok flowers.

The range of floral structure: flowers and fruits

In this section, a range of flowers and their fruits are shown.

Flower structure can be described in three ways:
1   As a half flower.
2   As a floral formula.
3   As a floral diagram.

The definitions of the terms used in describing floral structure are given in the glossary.

1 A half flower is a drawing of the inside of the flower cut in half longitudinally through the line between the axis of the stem and the bract of the flower, i.e. a vertical section through the floral diagram. The half drawn is the right half of the flower, viewed from left to right. Cut edges of petals and sepals are drawn as double lines.

2 A floral formula is a coded description of a flower, stating whether it is hermaphrodite, male, female or neuter, whether it is actinomorphic or zygomorphic, and the number of sepals, petals, stamens and carpels.

The symbols used are:
  P = perianth
  K = calyx (number of sepals)
  C = corolla (number of petals)
  A = androecium (number of stamens)
  G = gynaecium (number of carpels)
  $\oplus$ = actinomorphic (radially symmetrical)
  $+$ = zygomorphic (bilaterally symmetrical) sometimes
  ∞ = infinity (numerous, i.e. more than twelve)
  $\male$ = male
  $\female$ = female
  $\hermaphrodite$ = hermaphrodite, fomerly $\male\female$

A bracket indicates fused parts, e.g. K(5) means five fused sepals.
A line above or below the carpel number indicates whether the ovary is superior or inferior, e.g. $G\overline{1}$ means an inferior ovary and $G\underline{1}$ means a superior ovary, each of one carpel.
A linking line between two parts means that they are joined to each other, e.g. $\overline{P6\ A6}$ means that there are six stamens attached to the perianth lobes. This may also be written with brackets {P6 A6}

Some examples of floral formulae are:

(i) $\hermaphrodite \oplus$ K5 C5 A∞G∞
This means a hermaphrodite actinomorphic flower with five free sepals, five free petals, many stamens, and a superior apocarpous gynaecium of many free carpels.

(ii) $\hermaphrodite + $ K(5) {C(5) A4} G($\overline{2}$)
This means a hermaphrodite zygomorphic flower with five fused sepals, five fused petals, four stamens attached to the petals, and an inferior syncarpous gynaecium of two fused carpels.

3 A floral diagram is a formal plan of a flower. By convention it is not usually labelled (unless an unusual structure is shown) but each whorl of structures has a symbol (see next page).

Division	Spermatophyta	
Subdivision	Angiosperms	

Flower structure
Floral diagrams
Forms of receptacle

Floral diagram of generalised dicotyledon with most of flower parts free

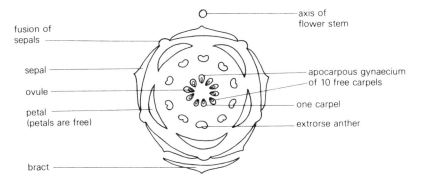

fusion of sepals

axis of flower stem

sepal

apocarpous gynaecium of 10 free carpels

ovule

one carpel

petal (petals are free)

extrorse anther

bract

Floral diagram of generalised monocotyledon with some fusion of flower parts

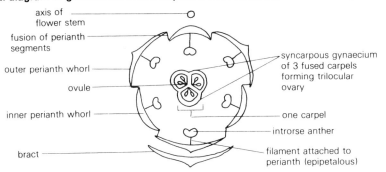

axis of flower stem

fusion of perianth segments

syncarpous gynaecium of 3 fused carpels forming trilocular ovary

outer perianth whorl

ovule

inner perianth whorl

one carpel

introrse anther

bract

filament attached to perianth (epipetalous)

Forms of receptacle in hypogynous, perigynous and epigynous flowers

Flowers are shown in L.S.

The receptacle is the structure on which flower parts arise.

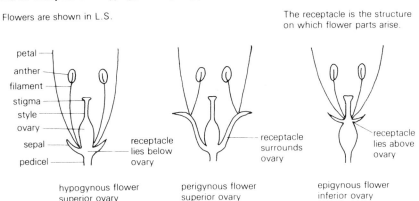

petal
anther
filament
stigma
style
ovary
sepal
pedicel

receptacle lies below ovary

receptacle surrounds ovary

receptacle lies above ovary

hypogynous flower
superior ovary

perigynous flower
superior ovary

epigynous flower
inferior ovary

Ranunculus species	Division	Spermatophyta
buttercup	Subdivision	Angiosperms
	Class	Dicotyledons
	Family	Ranunculaceae
flower and fruit	Genus	*Ranunculus*

Half flower of buttercup

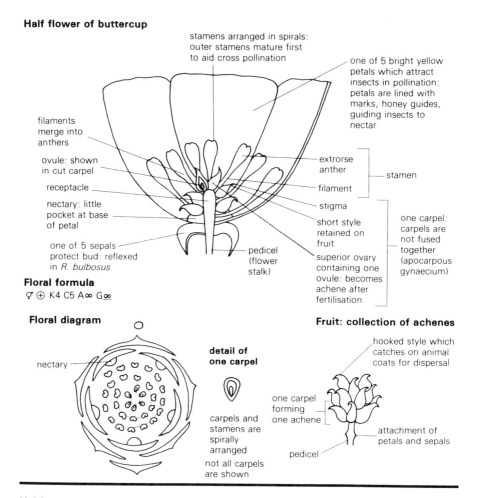

stamens arranged in spirals: outer stamens mature first to aid cross pollination

one of 5 bright yellow petals which attract insects in pollination: petals are lined with marks, honey guides, guiding insects to nectar

filaments merge into anthers

ovule: shown in cut carpel

receptacle

nectary: little pocket at base of petal

one of 5 sepals protect bud: reflexed in *R. bulbosus*

extrorse anther

filament

stigma

short style retained on fruit

superior ovary containing one ovule: becomes achene after fertilisation

pedicel (flower stalk)

stamen

one carpel: carpels are not fused together (apocarpous gynaecium)

Floral formula

♀ ⊕ K4 C5 A∞ G∞

Floral diagram

nectary

detail of one carpel

carpels and stamens are spirally arranged

not all carpels are shown

Fruit: collection of achenes

hooked style which catches on animal coats for dispersal

one carpel forming one achene

attachment of petals and sepals

pedicel

Habitat: grassland.

Pollination: entomophilous and unspecialised; pollinated by various insects especially small bees and flies. Protogynous, with extrorse anthers to reduce self pollination, but buttercups are often self sterile.

Fruit: is a collection of achenes, which are animal-dispersed or fall to the ground.

Notes: there are three common species of buttercup which flower in May. *Ranunculus bulbosus* is the earliest flowering, has reflexed sepals, and lives on drier grassland than the other two. *R. repens* has a furrowed flower stalk and *R. acris* a smooth one.

The family Ranunculaceae also includes clematis, anemone and delphinium; it is considered primitive due to the free parts, spiral arrangement, and unspecialised pollination.

Division	Spermatophyta
Subdivision	Angiosperms
Class	Dicotyledons
Family	Cruciferae
Genus	Cheiranthus

Cheiranthus cheiri
wallflower

flower and fruit

Half flower of wallflower

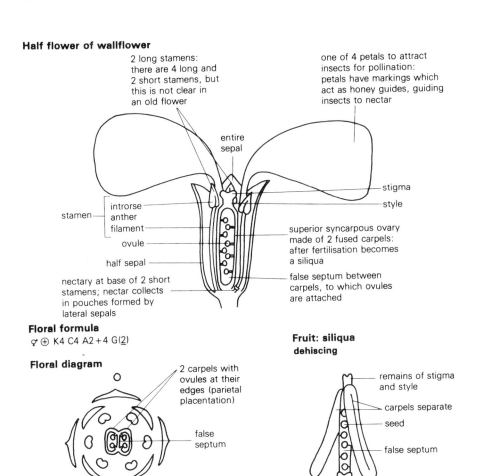

2 long stamens:
there are 4 long and
2 short stamens, but
this is not clear in
an old flower

one of 4 petals to attract
insects for pollination:
petals have markings which
act as honey guides, guiding
insects to nectar

entire
sepal

stigma

style

stamen —
introrse
anther

filament

ovule

half sepal

nectary at base of 2 short
stamens; nectar collects
in pouches formed by
lateral sepals

superior syncarpous ovary
made of 2 fused carpels:
after fertilisation becomes
a siliqua

false septum between
carpels, to which ovules
are attached

Floral formula
♀ ⊕ K4 C4 A2 + 4 G(2)

Floral diagram

2 carpels with
ovules at their
edges (parietal
placentation)

false
septum

**Fruit: siliqua
dehiscing**

remains of stigma
and style

carpels separate

seed

false septum

Habitat and distribution: wallflower is native to the eastern Mediterranean, but is widely naturalised and grows wild in lowland Britain, especially on walls.

Pollination: entomophilous by various bees and hover flies attracted by scent and colour; homogamous.

Fruit: is a siliqua which is dispersed by separation of the two carpels; the seeds, attached to the false septum, are blown out by wind.

Notes: the family Cruciferae (crucifers) is very easy to recognise by its cross-shaped flowers (Cruciferae means ''cross-bearing''). Many members are food plants, e.g. mustard, cress, and the brassicas such as cabbage, Brussels sprout, cauliflower, turnip, swede.

	Division	Spermatophyta
Fragaria strawberry **flower and fruit**	Subdivision	Angiosperms
	Class	Dicotyledons
	Family	Rosaceae
	Genus	*Fragaria*

Half flower of strawberry

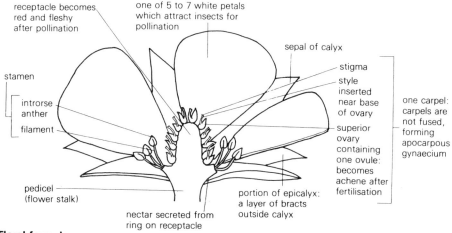

receptacle becomes red and fleshy after pollination

one of 5 to 7 white petals which attract insects for pollination

sepal of calyx

stamen

introrse anther

filament

stigma

style inserted near base of ovary

superior ovary containing one ovule: becomes achene after fertilisation

one carpel: carpels are not fused, forming apocarpous gynaecium

pedicel (flower stalk)

nectar secreted from ring on receptacle

portion of epicalyx: a layer of bracts outside calyx

Floral formula

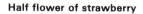

♀ ⊕ K5 C5 A20 G∞

The cultivated strawberry may have up to 7 petals and sepals.

Floral diagram

epicalyx

detail of one carpel

carpel wall

ovule

swollen receptacle

remains of stamen

remains of sepal

Fruit: false fruit, pseudocarp half fruit

remains of stigma and style

achene developed from ovary: passes through animal gut

red fleshy receptacle (torus) attracts birds for dispersal in the wild

This is a false fruit because the fleshy part is not the pericarp, but here is the receptacle.

Habitat: the wild strawberry, *Fragaria vesca*, is a woodland herb. Our cultivated strawberry, *Fragaria x ananassa*, is a hybrid between two American species and was produced in cultivation.

Pollination: unspecialised, by various insects. Protogynous to avoid selfing.

Fruit: is a group of achenes on an enlarged, red receptacle and is called a pseudocarp.

Notes: cultivated strawberries are usually produced from runners, not seeds, but alpine and Hautbois strawberries can be grown from seed.

Division	Spermatophyta
Subdivision	Angiosperms
Class	Dicotyledons
Family	Rosaceae
Genus	Prunus

Prunus avium
sweet cherry

flower and fruit

Half flower of sweet cherry

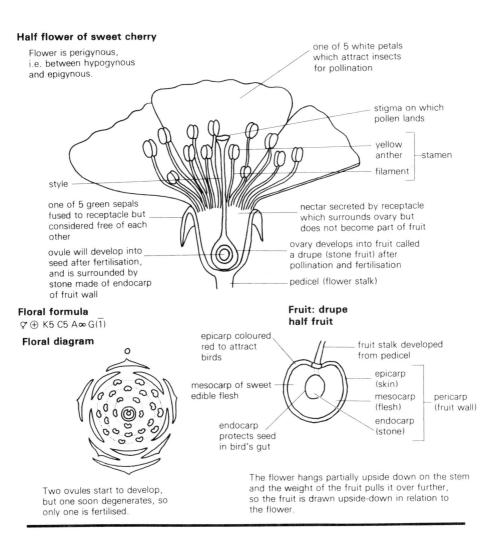

Flower is perigynous, i.e. between hypogynous and epigynous.

one of 5 white petals which attract insects for pollination

stigma on which pollen lands

yellow anther —stamen

filament

style

one of 5 green sepals fused to receptacle but considered free of each other

nectar secreted by receptacle which surrounds ovary but does not become part of fruit

ovule will develop into seed after fertilisation, and is surrounded by stone made of endocarp of fruit wall

ovary develops into fruit called a drupe (stone fruit) after pollination and fertilisation

pedicel (flower stalk)

Floral formula

$\male \oplus$ K5 C5 A∞ G($\overline{1}$)

Floral diagram

Two ovules start to develop, but one soon degenerates, so only one is fertilised.

Fruit: drupe
half fruit

epicarp coloured red to attract birds

fruit stalk developed from pedicel

mesocarp of sweet edible flesh

epicarp (skin)

mesocarp (flesh)

endocarp (stone)

pericarp (fruit wall)

endocarp protects seed in bird's gut

The flower hangs partially upside down on the stem and the weight of the fruit pulls it over further, so the fruit is drawn upside-down in relation to the flower.

Habitat: *Prunus avium*, in its wild form, is native to Britain in woods and hedgerows.
Pollination: entomophilous by various insects. Homogamous, but all varieties are completely self sterile.
Fruit: is a drupe (stone fruit) dispersed by being eaten by birds and the seed voided; attracted by red colour of epicarp and sweet mesocarp.
Notes: *Prunus avium* is the ancestor of our sweet cherries. The sour cherry is *Prunus cerasus*. The genus *Prunus* also includes plums, apricots and almonds. Flowering cherries usually have many layers of petals and are often sterile.

Malus apple flower and fruit	Division	Spermatophyta
	Subdivision	Angiosperms
	Class	Dicotyledons
	Family	Rosaceae
	Genus	*Malus*

Half flower of apple

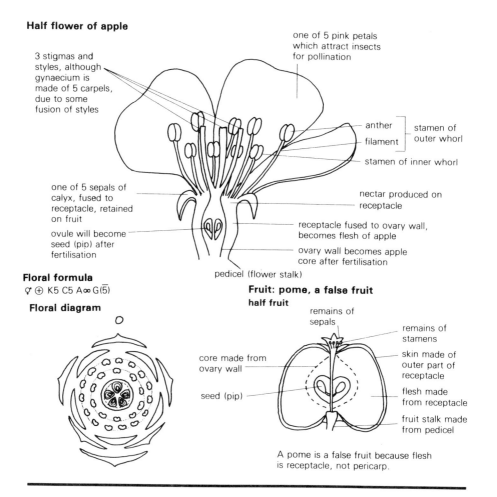

3 stigmas and styles, although gynaecium is made of 5 carpels, due to some fusion of styles

one of 5 pink petals which attract insects for pollination

anther
filament
— stamen of outer whorl

stamen of inner whorl

one of 5 sepals of calyx, fused to receptacle, retained on fruit

nectar produced on receptacle

ovule will become seed (pip) after fertilisation

receptacle fused to ovary wall, becomes flesh of apple

ovary wall becomes apple core after fertilisation

pedicel (flower stalk)

Floral formula

$\female \oplus K5\ C5\ A\infty\ G\overline{(5)}$

Floral diagram

Fruit: pome, a false fruit
half fruit

remains of sepals

remains of stamens

core made from ovary wall

skin made of outer part of receptacle

seed (pip)

flesh made from receptacle

fruit stalk made from pedicel

A pome is a false fruit because flesh is receptacle, not pericarp.

Habitat: the wild crab apple, *Malus pumila,* is native to Britain in woods and hedges. Modern varieties were produced mainly from this species by selection and breeding.

Pollination: entomophilous, mainly by bees and long-tongued flies. Slightly protogynous, so stigmas are usually pollinated from another flower before stamens are ripe. Most varieties are self sterile. but self pollination is sometimes possible.

Fruit: is a pome, which in the wild crab apple is dispersed by birds, attracted by its red colour and sweet taste.

Notes: cultivated varieties are propagated by grafting since seeds produce variable plants, similar to wild crab apples, because of cross pollination.

The pear is also a pome but has a different taste and texture due to grit cells in its flesh and different acids and sugars.

Division	Spermatophyta
Subdivision	Angiosperms
Class	Dicotyledons
Family	Rosaceae
Genus	Crataegus

Crataegus monogyna
hawthorn

flower and fruit

Half flower of hawthorn

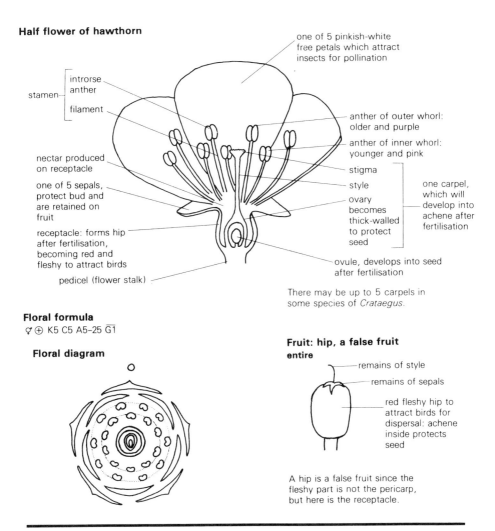

one of 5 pinkish-white free petals which attract insects for pollination

stamen
- introrse anther
- filament

anther of outer whorl: older and purple

anther of inner whorl: younger and pink

nectar produced on receptacle

stigma

one of 5 sepals, protect bud and are retained on fruit

style

ovary becomes thick-walled to protect seed

one carpel, which will develop into achene after fertilisation

receptacle: forms hip after fertilisation, becoming red and fleshy to attract birds

ovule, develops into seed after fertilisation

pedicel (flower stalk)

There may be up to 5 carpels in some species of *Crataegus*.

Floral formula

♀ ⊕ K5 C5 A5–25 $\overline{G1}$

Floral diagram

Fruit: hip, a false fruit
entire

remains of style

remains of sepals

red fleshy hip to attract birds for dispersal: achene inside protects seed

A hip is a false fruit since the fleshy part is not the pericarp, but here is the receptacle.

Habitat: scrub, woods, hedges.
Pollination: entomophilous by flies, bees, wasps and beetles attracted by the scent and colour of petals and stamens. Protogynous to avoid selfing.
Fruit: is a hip, dispersed by birds which eat the fleshy receptacles and either void or do not swallow the achenes; these have thick coats to avoid digestion.
Notes: there are two species of *Crataegus* in Britain. *C. monogyna*, hawthorn, has one style and one ovule, while *C. oxyacanthoides*, Midland hawthorn, has two styles and two ovules, but there are hybrids with varying numbers. The pink and double-flowered hawthorns are forms of our native species.

Laburnum anagyroides	Division	Spermatophyta
laburnum	Subdivision	Angiosperms
	Class	Dicotyledons
	Family	Leguminosae
flower and fruit	Genus	Laburnum

Half flower of laburnum

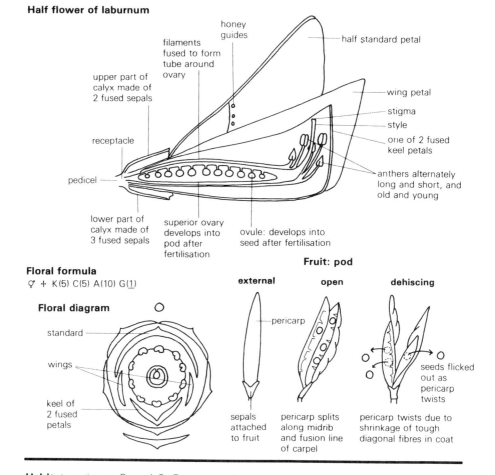

Floral formula

♀ + K(5) C(5) A(10) G(1)

Floral diagram

Fruit: pod

Habitat: native to S. and C. Europe; widely planted, and naturalised in waste places.

Pollination: entomophilous by large bees. Some legumes produce nectar, e.g. clover, but many do not. Insects land on the keel, and the stigma and anthers spring out and dust the insect's underside with pollen. (In broom this is so vigorous that the top of the insect's abdomen is also dusted.) Many legumes can self pollinate, but some show self incompatibility.

Fruit: is a pod, dispersed by an explosive mechanism. Laburnum seeds are poisonous to man. The pods remain on the tree all winter.

Notes: the family Leguminosae (legumes) includes peas, beans, clovers, lucerne, gorse, lupin, broom, etc. and also the peanut, whose ''nuts'' are seeds. In all members of the family the fruit is a pod, also called a legume, and the roots contain the nitrogen-fixing bacterium, *Rhizobium,* in root nodules.

Division	Spermatophyta
Subdivision	Angiosperms
Class	Dicotyledons
Family	Labiatae
Genus	Lamium

Lamium album
white dead-nettle

flower and fruit

Half flower of white deadnettle

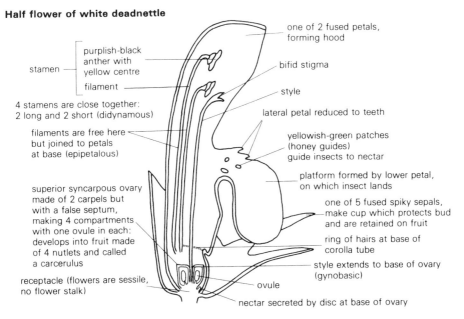

- one of 2 fused petals, forming hood
- bifid stigma
- style
- lateral petal reduced to teeth
- yellowish-green patches (honey guides) guide insects to nectar
- platform formed by lower petal, on which insect lands
- one of 5 fused spiky sepals, make cup which protects bud and are retained on fruit
- ring of hairs at base of corolla tube
- style extends to base of ovary (gynobasic)
- nectar secreted by disc at base of ovary
- ovule

stamen
- purplish-black anther with yellow centre
- filament

4 stamens are close together: 2 long and 2 short (didynamous)

filaments are free here but joined to petals at base (epipetalous)

superior syncarpous ovary made of 2 carpels but with a false septum, making 4 compartments with one ovule in each: develops into fruit made of 4 nutlets and called a carcerulus

receptacle (flowers are sessile, no flower stalk)

Floral formula

$♀ + K(5) \{C(5) A4\} G(\underline{2})$

Floral diagram   ○

Fruit: carcerulus

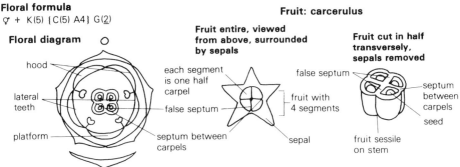

Fruit entire, viewed from above, surrounded by sepals

Fruit cut in half transversely, sepals removed

- hood
- lateral teeth
- platform
- septum between carpels
- each segment is one half carpel
- false septum
- fruit with 4 segments
- sepal
- false septum
- septum between carpels
- seed
- fruit sessile on stem

Habitat: hedge-banks, roadsides, waste places; native.

Pollination: entomophilous by long-tongued insects like bumble bees which alight on the lower petal and push their heads into the corolla tube so that their backs touch the anthers and stigma. Homogamous, and can self pollinate. Many labiates are protandrous; the anthers hang lower and the stigma lobes are closed, and do not open until anthers have shed pollen, encouraging crossing.

Fruit: is a carcerulus which breaks into four nutlets that are animal-dispersed.

Notes: the Labiatae (labiates) is a very natural family whose members typically have square stems and opposite leaves. It includes many cooking herbs with fragrant essential oils, e.g. mint, thyme, sage, rosemary, basil.

Lycopersicon esculentum	*Division*	Spermatophyta
tomato	*Subdivision*	Angiosperms
	Class	Dicotyledons
flower and fruit	*Family*	Solanaceae
	Genus	*Lycopersicon*

Half flower of tomato

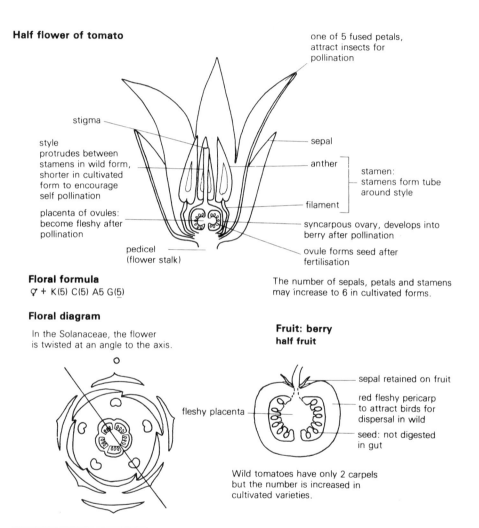

one of 5 fused petals,
attract insects for
pollination

stigma

style
protrudes between
stamens in wild form,
shorter in cultivated
form to encourage
self pollination

placenta of ovules:
become fleshy after
pollination

pedicel
(flower stalk)

sepal

anther

stamen:
stamens form tube
around style

filament

syncarpous ovary, develops into
berry after pollination

ovule forms seed after
fertilisation

Floral formula
♀ + K(5) C(5) A5 G(5)

The number of sepals, petals and stamens
may increase to 6 in cultivated forms.

Floral diagram

In the Solanaceae, the flower
is twisted at an angle to the axis.

**Fruit: berry
half fruit**

fleshy placenta

sepal retained on fruit

red fleshy pericarp
to attract birds for
dispersal in wild

seed: not digested
in gut

Wild tomatoes have only 2 carpels
but the number is increased in
cultivated varieties.

Range: native to lower Andes of South America; long cultivated there.
Pollination: in the wild, by insects, especially bees; the cultivated form is self
fertile, a characteristic developed in cultivation under glass, but flowers need to be
tapped in order for pollen to fall onto the stigma. If grown out of doors they are
insect-pollinated.
Fruit: is a berry, in nature dispersed by birds.
Notes: the family Solanaceae also includes sweet peppers, chilli peppers and the
aubergine (whose fruits are also edible berries), as well as the potato, deadly, and
woody nightshades, in which the berry and many other parts are poisonous.

Division	Spermatophyta
Subdivision	Angiosperms
Class	Dicotyledons
Family	Aceraceae
Genus	Acer

Acer pseudoplatanus
sycamore

flower and fruit

Young half flower: showing both stamens and ovary

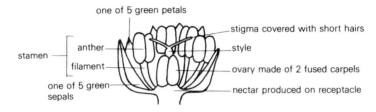

one of 5 green petals
stigma covered with short hairs
anther
style
stamen
filament
ovary made of 2 fused carpels
one of 5 green sepals
nectar produced on receptacle

Entire male flower

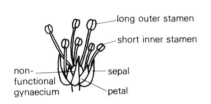

long outer stamen
short inner stamen
non-functional gynaecium
sepal
petal

Entire female flower

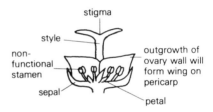

stigma
style
non-functional stamen
outgrowth of ovary wall will form wing on pericarp
sepal
petal

Floral formula
♀ ⊕ K5 C5 A4+4 G(2)

Floral diagram

Fruit: double samara entire

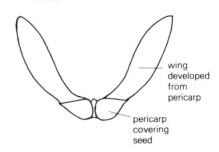

wing developed from pericarp
pericarp covering seed

This fruit is sometimes called a schizocarp because it splits into separate parts.

Habitat: common in woods, hedges, etc., throughout Britain. Not native, but introduced from central or southern Europe in the 15th or 16th century.
Pollination: both entomophilous and anemophilous. Flowers are borne in hanging branched racemes with slight scent and contain nectar to attract bees, but some pollen is wind-borne. Stamens and an ovary are present in all flowers, but some become functionally male and others functionally female as they mature.
Fruit: is a double samara and is wind dispersed, the wing acting as a helicopter blade. Single samaras are found in ash and elm.

Corylus avellana	Division	Spermatophyta
hazel	Subdivision	Angiosperms
	Class	Dicotyledons
flower and fruit	Family	Corylaceae
	Genus	*Corylus*

Two male flowers

Floral formula

♂ + P0 A4 G0

Floral diagram

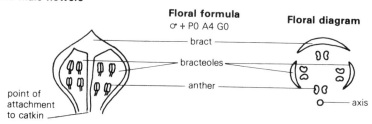

bract
bracteoles
anther
point of attachment to catkin
axis

Two female flowers

Floral formula

♀ + P4 A0 G(2̄)

Floral diagram

(1) and (2) are two female flowers

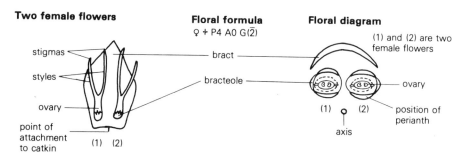

stigmas
styles
ovary
point of attachment to catkin
(1) (2)
bract
bracteole
ovary
position of perianth
axis

Male and female catkins

Fruit: true nut entire

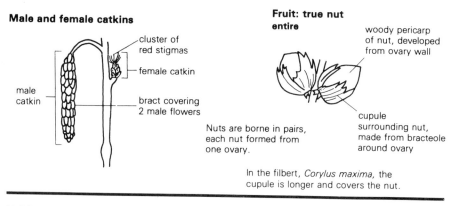

cluster of red stigmas
female catkin
male catkin
bract covering 2 male flowers

Nuts are borne in pairs, each nut formed from one ovary.

woody pericarp of nut, developed from ovary wall

cupule surrounding nut, made from bracteole around ovary

In the filbert, *Corylus maxima*, the cupule is longer and covers the nut.

Habitat: native to Britain in woods, scrub, hedges.

Pollination: anemophilous. Male catkins are easily blown in wind, dislodging pollen, while stigmas of female catkins protrude from flowers to receive it.

Fruit: is a true nut, surrounded by a cupule made from bracteoles of female flower. It is dispersed by mammals which bury nuts as a food store, then do not find them all.

Notes: flowers are borne in male and female catkins. Individual flowers are very reduced, with no perianth in the male, and a slight perianth in the female. Within the catkins flowers are borne in pairs, each pair surrounded by a common bract and each flower surrounded by its own bracteole.

114

Division	Spermatophyta
Subdivision	Angiosperms
Class	Dicotyledons
Family	Compositae

Dandelion, daisy, thistle

structure of inflorescence in Compositae

L.S. capitulum of dandelion: all florets ligulate

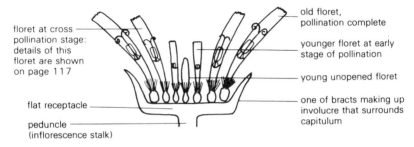

floret at cross pollination stage: details of this floret are shown on page 117

flat receptacle

peduncle (inflorescence stalk)

old floret, pollination complete

younger floret at early stage of pollination

young unopened floret

one of bracts making up involucre that surrounds capitulum

L.S. capitulum of daisy: outer florets ligulate, inner tubular

Outer florets are called ray florets and inner florets are called disc florets in an inflorescence with two types of floret.

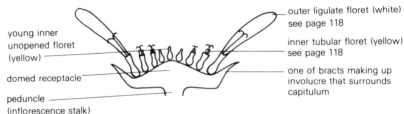

young inner unopened floret (yellow)

domed receptacle

peduncle (inflorescence stalk)

outer ligulate floret (white) see page 118

inner tubular floret (yellow) see page 118

one of bracts making up involucre that surrounds capitulum

L.S. capitulum of thistle: all florets tubular

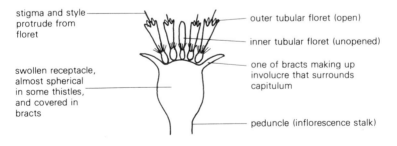

stigma and style protrude from floret

swollen receptacle, almost spherical in some thistles, and covered in bracts

outer tubular floret (open)

inner tubular floret (unopened)

one of bracts making up involucre that surrounds capitulum

peduncle (inflorescence stalk)

The inflorescence in the Compositae is the apparent "flower", and is called a **capitulum.** The actual flowers are tiny and each is called a **floret.** There are two types of floret; **ligulate florets** have the corolla expanded on one side to form a strap-shaped structure, and **tubular florets** have all petals the same size forming a tube.
Notes: the Compositae (composites) is a very large and successful family. It includes many "weeds" such as dandelion, daisy, thistles, ragworts, groundsel, and also sunflower, chrysanthemum, marigold, chicory and lettuce.

Pollination in Compositae	Division	Spermatophyta
	Subdivision	Angiosperms
	Class	Dicotyledons
	Family	Compositae

L.S. floret of Compositae drawn diagrammatically to show pollination

1 Young floret

2 As floret grows older, stigma elongates beyond anthers

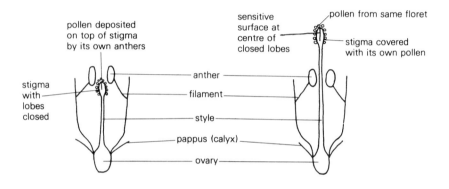

3 Stigma opens and cross pollination by insects occurs at this stage

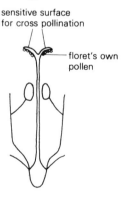

4 Old floret: if cross pollination fails, stigma curls round so that self pollination is possible

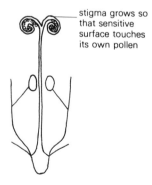

Pollination in most Compositae is by the same mechanism. They are entomophilous. Insects are attracted by the colour of the whole inflorescence, and crawl over the surface seeking nectar at the base of the style, and depositing or collecting pollen. Self pollination occurs if crossing fails, as shown in the diagrams.

The ovary initially consists of two fused carpels, but during development only one carpel persists, containing one ovule.

Division	Spermatophyta
Subdivision	Angiosperms
Class	Dicotyledons
Family	Compositae
Genus	Taraxacum

Taraxacum officinale
dandelion

flower and fruit

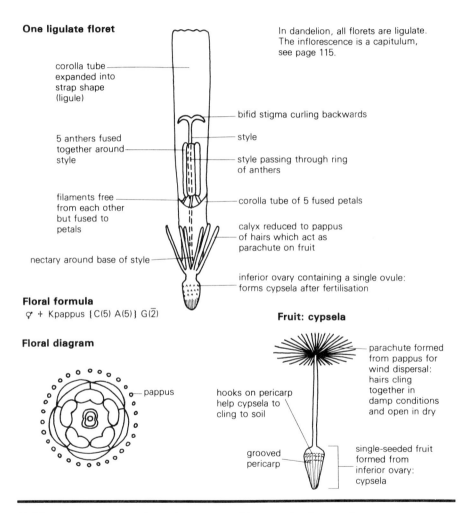

One ligulate floret

In dandelion, all florets are ligulate. The inflorescence is a capitulum, see page 115.

corolla tube expanded into strap shape (ligule)

bifid stigma curling backwards

style

5 anthers fused together around style

style passing through ring of anthers

filaments free from each other but fused to petals

corolla tube of 5 fused petals

calyx reduced to pappus of hairs which act as parachute on fruit

nectary around base of style

inferior ovary containing a single ovule: forms cypsela after fertilisation

Floral formula

$\male$ + Kpappus {C(5) A(5)} G($\overline{2}$)

Fruit: cypsela

Floral diagram

pappus

parachute formed from pappus for wind dispersal: hairs cling together in damp conditions and open in dry

hooks on pericarp help cypsela to cling to soil

grooved pericarp

single-seeded fruit formed from inferior ovary: cypsela

Habitat: pastures, meadows, lawns, waysides, waste places, etc.

Pollination: although dandelions appear to have the normal pollination of the Compositae (see page 116), many are apomictic, i.e. produce fruit without fertilisation.

Fruit: is a cypsela with a parachute of hairs for wind dispersal.

Notes: dandelion is well adapted to grazing and mowing. It has a basal rosette of leaves which is pressed downwards due to more rapid growth of the upper side of each leaf. The root is contractile, ensuring that the rosette remains at ground level, and the apical meristem lies in the short rootstock from which leaves arise. All of these adaptations keep it flat to avoid predator's teeth or lawn mower blades. The fruits can develop even when the flower head is cut off the plant.

Bellis perennis	_Division_ Spermatophyta
daisy	_Subdivision_ Angiosperms
	Class Dicotyledons
	Family Compositae
flower and fruit	_Genus_ _Bellis_

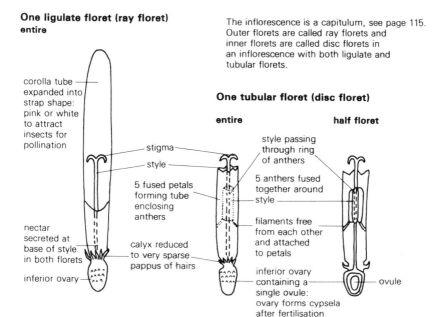

One ligulate floret (ray floret)
entire

The inflorescence is a capitulum, see page 115. Outer florets are called ray florets and inner florets are called disc florets in an inflorescence with both ligulate and tubular florets.

corolla tube expanded into strap shape: pink or white to attract insects for pollination

stigma

style

5 fused petals forming tube enclosing anthers

nectar secreted at base of style in both florets

calyx reduced to very sparse pappus of hairs

inferior ovary

One tubular floret (disc floret)

entire   half floret

style passing through ring of anthers

5 anthers fused together around style

filaments free from each other and attached to petals

inferior ovary containing a single ovule: ovary forms cypsela after fertilisation in both florets

ovule

Floral formula
♀ + Kpappus C(5) A0 G($\overline{2}$)

Floral formula
♀ ⊕ Kpappus {C(5) A(5)} G($\overline{2}$)

Floral diagram
ligulate ray floret

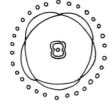

Floral diagram
tubular disc floret

Fruit: cypsela

sparse pappus not used in wind dispersal

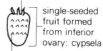

single-seeded fruit formed from inferior ovary: cypsela

hooks on pericarp help cypsela cling to soil

Habitat: abundant in short grassland.
Pollination: see pollination in Compositae (page 116).
Fruit: is a cypsela but has almost no pappus, so it is dispersed just around the parent, but it can be carried further by ants, or in mud on animals' feet.
Notes: daisy is well adapted to grazing and mowing like the dandelion (see page 117). In addition, short prostrate shoots develop, so the plant spreads in patches.

Division	Spermatophyta
Subdivision	Angiosperms
Class	Monocotyledons
Family	Liliaceae
Genus	Endymion

Endymion non-scriptus
bluebell

flower and fruit

Half flower of bluebell

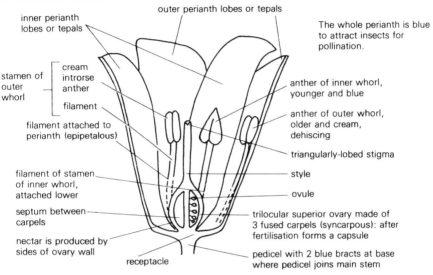

- inner perianth lobes or tepals
- outer perianth lobes or tepals
- The whole perianth is blue to attract insects for pollination.
- stamen of outer whorl
  - cream introrse anther
  - filament
- filament attached to perianth (epipetalous)
- anther of inner whorl, younger and blue
- anther of outer whorl, older and cream, dehiscing
- triangularly-lobed stigma
- filament of stamen of inner whorl, attached lower
- style
- ovule
- septum between carpels
- nectar is produced by sides of ovary wall
- receptacle
- trilocular superior ovary made of 3 fused carpels (syncarpous): after fertilisation forms a capsule
- pedicel with 2 blue bracts at base where pedicel joins main stem

In the half flower it is usual to draw 3 whole stamens rather than 2 whole and 2 half.

Floral formula

$\male\female \oplus \{P3+3\ A3+3\}\ G(\underline{3})$

Floral diagram

Fruit: capsule
entire, dehiscing

- opening through which seeds are dispersed
- fruit splits open at the place where carpels are joined
- 3 fused carpels

Habitat: common in woods, hedge-banks, etc.
Pollination: entomophilous by a variety of insects, especially bees.
Fruit: is a capsule which opens at the top so that the seeds can be shaken out in the wind.
Notes: the family Liliaceae includes the lily, tulip, lily of the valley, hyacinth, asparagus, and sometimes onions, although these may be placed in a separate family, the Alliaceae. Not all members are epipetalous, or have capsules as fruits. In some genera the fruit is a berry and/or the perianth lobes are fused. Vegetative propagation by bulbs is common.

Generalised grass	Division	Spermatophyta
rye grass and oat as examples	Subdivision	Angiosperms
	Class	Monocotyledons
vegetative and inflorescence structure	Family	Gramineae

Vegetative features

Inflorescences

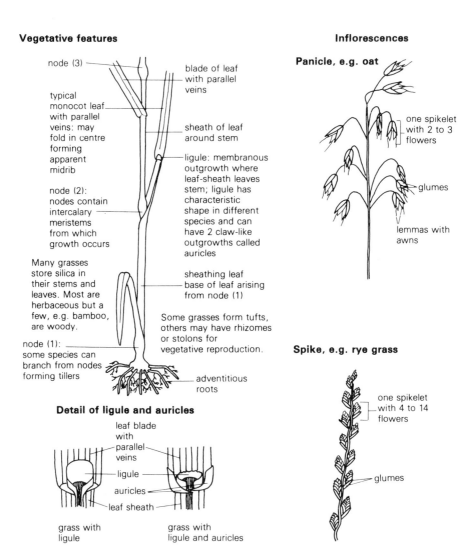

node (3)

typical monocot leaf with parallel veins: may fold in centre forming apparent midrib

node (2): nodes contain intercalary meristems from which growth occurs

Many grasses store silica in their stems and leaves. Most are herbaceous but a few, e.g. bamboo, are woody.

node (1): some species can branch from nodes forming tillers

blade of leaf with parallel veins

sheath of leaf around stem

ligule: membranous outgrowth where leaf-sheath leaves stem; ligule has characteristic shape in different species and can have 2 claw-like outgrowths called auricles

sheathing leaf base of leaf arising from node (1)

Some grasses form tufts, others may have rhizomes or stolons for vegetative reproduction.

adventitious roots

Panicle, e.g. oat

one spikelet with 2 to 3 flowers

glumes

lemmas with awns

Spike, e.g. rye grass

one spikelet with 4 to 14 flowers

glumes

Detail of ligule and auricles

leaf blade with parallel veins

ligule

auricles

leaf sheath

grass with ligule

grass with ligule and auricles

Grasses are thought to be the most advanced family of monocots. They are well adapted to survive grazing by intercalary meristems near the nodes, which enable them to grow from the base, so it does not matter if the top is bitten off. Grazing and trampling also encourages tillers to form.

The inflorescence is usually either a **spike** or **panicle**. It is made up of small units called **spikelets**, see next page.

The term ''grass'' is the common name for the family Gramineae.

Division	Spermatophyta
Subdivision	Angiosperms
Class	Monocotyledons
Family	Gramineae

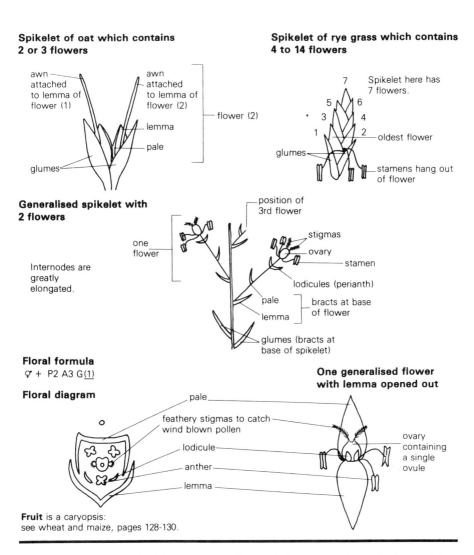

**Spikelet of oat which contains
2 or 3 flowers**

awn
attached
to lemma of
flower (1)

awn
attached
to lemma of
flower (2)

flower (2)

lemma

pale

glumes

**Spikelet of rye grass which contains
4 to 14 flowers**

7   Spikelet here has
     7 flowers.
5   6
3   4
1   2   oldest flower
glumes

stamens hang out
of flower

**Generalised spikelet with
2 flowers**

one
flower

Internodes are
greatly
elongated.

position of
3rd flower

stigmas

ovary

stamen

lodicules (perianth)

pale

lemma

bracts at base
of flower

glumes (bracts at
base of spikelet)

Floral formula
♀ + P2 A3 G(1)

Floral diagram

**One generalised flower
with lemma opened out**

pale

feathery stigmas to catch
wind blown pollen

lodicule

anther

lemma

ovary
containing
a single
ovule

Fruit is a caryopsis:
see wheat and maize, pages 128-130.

A **spikelet** is the unit of which any type of grass inflorescence is made. It consists of several flowers surrounded by two bracts called **glumes.** Each flower is also surrounded by two bracts, an outer **lemma** and an inner **pale.** In some grasses, e.g. oat, the lemma, pale, or glumes have bristles called **awns.** The perianth is reduced to two **lodicules.**
Pollination: anemophilous. Adaptations include feathery stigmas to catch pollen and anthers hanging out of the flower. Pollen is light and dusty.
Fruit dispersal: may be by wind, or by clinging to animal coats.

Cocos nucifera	*Division*	Spermatophyta
coconut palm	*Subdivision*	Angiosperms
	Class	Monocotyledons
	Family	Palmae
palm and fruit	*Genus*	*Cocos*

Entire palm

Detail of generalised palm stem to show leaf bases and formation of trunk

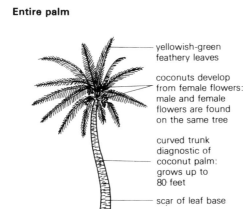

yellowish-green feathery leaves

coconuts develop from female flowers: male and female flowers are found on the same tree

curved trunk diagnostic of coconut palm: grows up to 80 feet

scar of leaf base

trunk thicker at base

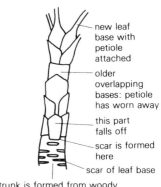

new leaf base with petiole attached

older overlapping bases: petiole has worn away

this part falls off

scar is formed here

scar of leaf base

trunk is formed from woody remains of leaf bases

L.S. through fruit

solid endosperm: white "flesh" or "meat" called copra

liquid endosperm: coconut milk

3 "eyes" of softer tissue here, through which plumule and radicle emerge

epicarp: outer skin, green when unripe, brown when ripe

mesocarp: fibrous coir which is removed before export and used in coconut matting, etc.; it holds air and makes coconut buoyant for sea dispersal

endocarp: shell around seed; the outer shell when coconut is exported, and to which remains of fibrous mesocarp are still attached

testa of seed

embryo

The coconut is a drupe: the epicarp, mesocarp and endocarp make up the pericarp.

Distribution: coconut palms are widespread in the tropics, especially at sea level.
Notes: coconut fruits are dispersed by ocean currents and will float for a long time. They require a sandy soil to germinate, and can tolerate salt, so they can grow on sandy beaches inhospitable to many plants. Since they are so widespread, there is doubt about their original homeland.

Other palms include date, sago and sugar palms. The leaves of palms may be feather- or fan-shaped. The trunks are not true wood (which is not made in monocotyledons), but are the hardened bases of the leaves.

Seeds and germination

L.S. generalised endospermic seed of dicotyledon

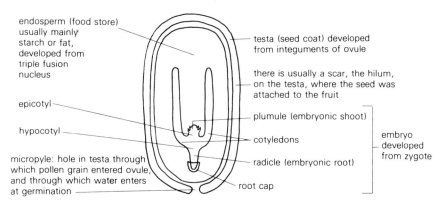

endosperm (food store) usually mainly starch or fat, developed from triple fusion nucleus

testa (seed coat) developed from integuments of ovule

there is usually a scar, the hilum, on the testa, where the seed was attached to the fruit

epicotyl

plumule (embryonic shoot)

hypocotyl

cotyledons

micropyle: hole in testa through which pollen grain entered ovule, and through which water enters at germination

radicle (embryonic root)

root cap

embryo developed from zygote

L.S. generalised non-endospermic seed of dicotyledon (not to same scale)

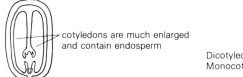

cotyledons are much enlarged and contain endosperm

Dicotyledons have 2 cotyledons.
Monocotyledons have one cotyledon.

A **seed** is a reproductive and dispersive structure found in spermatophytes only, and consists of a testa (seed coat), endosperm (food store), and an embryo made of a radicle, plumule, and cotyledons.

Differences between the seeds of gymnosperms and angiosperms

Gymnosperms	Angiosperms
Testa made of 1 integument.	Testa made of 2 integuments.
Endosperm haploid and made from female prothallus.	Endosperm triploid and made from triple fusion nucleus.
Embryo consists of radicle, plumule and varying numbers of cotyledons.	Embryo consists of radicle, plumule and 1 or 2 cotyledons.
Seeds develop naked on megasporophylls.	Seeds develop in fruit formed from fused megasporophylls (carpels).

A **non-endospermic** seed is one in which the endosperm is absorbed into the cotyledons, which often become thick and fleshy, e.g. pea, beans, sunflower.
An **endospermic seed** is one in which the endosperm is not absorbed into the cotyledons, e.g. cereals, buttercup.
In some seeds the food store is made of nucellus and is called **perisperm.**
Some apparent ''seeds'' are one-seeded fruits, e.g. the cypsela and caryopsis.

Germination

Germination is the collective name for those processes occurring between the time that the seed takes in water to when it opens its cotyledons or first leaves and begins to photosynthesise.

A **dormant** seed is one which is not germinating but is still **viable** (alive). The length of dormancy before death varies: willow seeds are viable for only one day, but viable seeds of *Nelumbo nucifera*, the Indian lotus, have been found that are known to have been buried for 160 years, and probably for 250, and which have been radiocarbon dated to 1000 years old. Most common seeds survive for from five to ten years, but vacuum packing, and dry and cold conditions can prolong viability.

The conditions necessary for germination

All seeds require:
(a) Water: to provide a medium in which enzymes can dissolve and in which chemical reactions can occur, and to allow hydrolysis of food stores.
(b) Oxygen: for respiration.
(c) Suitable temperature: for enzyme activity.
(d) Absence of toxic conditions, e.g. of carbon dioxide: for enzyme activity.

Some seeds also require one or more of the following:
(e) A period of cold before germination: to avoid autumn germination.
(f) Breaking of the seed coat: in bird-dispersed seeds which have a thick coat to resist grinding in the bird gut.
(g) Light: in very small seeds which must be near the surface to germinate.
(h) Time: some seeds have an undeveloped embryo which grows after dispersal.

The process of germination

Germination occurs when the necessary conditions have been met.

1 Water is absorbed through the micropyle and the seed swells considerably, rupturing the coat so that the radicle and plumule can emerge.

Enzymes become soluble and hydrolyse the insoluble food store to soluble products. Starch is converted by the enzyme amylase to maltose, which diffuses to the embryo where it is hydrolysed to glucose and used for respiration or growth. Fat stores may be hydrolysed to fatty acids and respired directly, or first converted to sugars via the glyoxalate cycle.

In cereals, enzymes are synthesized in the aleurone layer during germination, in many cases under control of the hormone, gibberellin.

2 The radicle breaks through the testa and grows downwards by positive geotropism, protected by the root cap.

3 The plumule grows upwards by negative geotropism at first, but once in the light, the stem grows by positive phototropism. In most seedlings the plumule is bent at first, the hypocotyl forming a plumule hook protecting the delicate meristem of the plumule as it forces its way through the soil. In grasses the coleoptile is protective and there is no plumule hook.

Tropisms are under the control of the hormone, auxin.

4 The cotyledons may be carried above the soil and become photosynthetic, forming the first ''seed leaves''; this is epigeal germination. If the cotyledons remain below soil, germination is hypogeal. The first leaves of a seedling are often different from the rest of the leaves, whether or not they are cotyledons.

Division	Spermatophyta
Subdivision	Angiosperms
Class	Dicotyledons
Family	Leguminosae
Genus	Phaseolus

Seed of French bean: non-endospermic seed

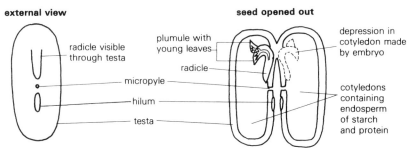

external view

seed opened out

radicle visible through testa

plumule with young leaves

depression in cotyledon made by embryo

radicle

micropyle

hilum

testa

cotyledons containing endosperm of starch and protein

Stages in germination: epigeal

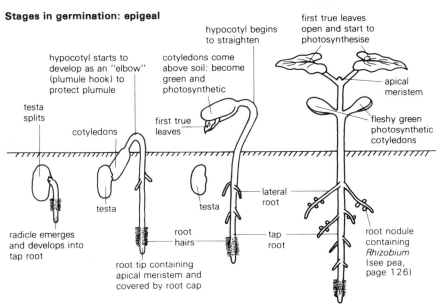

hypocotyl starts to develop as an "elbow" (plumule hook) to protect plumule

cotyledons come above soil: become green and photosynthetic

hypocotyl begins to straighten

first true leaves open and start to photosynthesise

testa splits

cotyledons

first true leaves

apical meristem

fleshy green photosynthetic cotyledons

radicle emerges and develops into tap root

testa

root hairs

testa

lateral root

tap root

root nodule containing *Rhizobium* (see pea, page 126)

root tip containing apical meristem and covered by root cap

Beans of the genus *Phaseolus* are native to the New World. The runner bean is *P. coccineus* and the kidney bean is the same species as the French bean (so called because it was introduced to Britain from France).

The broad bean is *Vicia faba* and is native to the Old World.

Runner beans and broad beans have hypogeal germination. The twining of a runner bean around a support is an autonomic movement, not a haptotropism as the pea tendril.

Beans and peas are pulses and store starch and protein, but the protein lacks the essential sulphur-containing amino acid, methionine. But it does contain lysine, deficient in cereal protein, so a suitable mixture of cereals and pulses can provide most of the essential amino acids.

Pisum sativum	Division	Spermatophyta
pea	Subdivision	Angiosperms
	Class	Dicotyledons
	Family	Leguminosae
seed and germination	Genus	Pisum

Seed of pea: non-endospermic seed

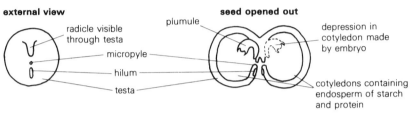

external view

radicle visible through testa
micropyle
hilum
testa

seed opened out

plumule

depression in cotyledon made by embryo

cotyledons containing endosperm of starch and protein

Stages in germination: hypogeal

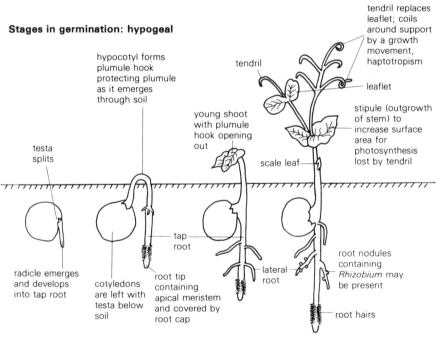

hypocotyl forms plumule hook protecting plumule as it emerges through soil

tendril

tendril replaces leaflet; coils around support by a growth movement, haptotropism

leaflet

young shoot with plumule hook opening out

stipule (outgrowth of stem) to increase surface area for photosynthesis lost by tendril

testa splits

scale leaf

tap root

radicle emerges and develops into tap root

cotyledons are left with testa below soil

root tip containing apical meristem and covered by root cap

lateral root

root nodules containing *Rhizobium* may be present

root hairs

Peas are probably native to the Near East, but have been cultivated in Europe for many centuries.

Peas, beans, and other legumes are called pulses, and contain similar food stores; for details see French bean (page 125).

The roots of legumes become infected with the nitrogen-fixing bacterium, *Rhizobium,* in the soil, forming root nodules. The bacterium forms a symbiotic association with the legume, enabling it to use nitrogen from the air, so that it does not draw on nitrates from the soil. This is why legumes are important in crop rotations. If grown in sterile, soil-less compounds nodules do not develop because *Rhizobium* is absent.

Division	Spermatophyta
Subdivision	Angiosperms
Class	Dicotyledons
Family	Compositae
Genus	Helianthus

Helianthus annuus
sunflower

seed and germination

Cypsela (fruit) of sunflower: non-endospermic seed

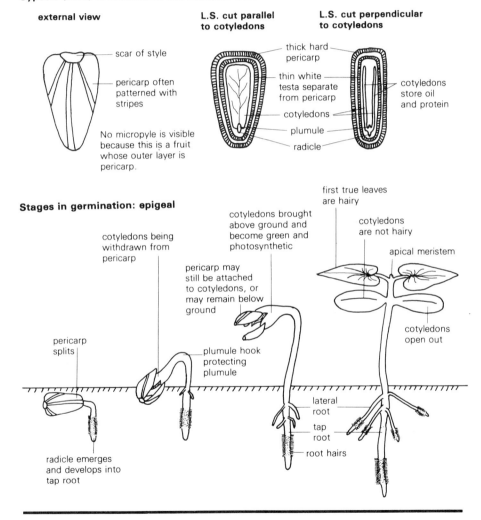

external view

L.S. cut parallel
to cotyledons

L.S. cut perpendicular
to cotyledons

scar of style

pericarp often
patterned with
stripes

No micropyle is visible
because this is a fruit
whose outer layer is
pericarp.

thick hard
pericarp

thin white
testa separate
from pericarp

cotyledons

plumule

radicle

cotyledons
store oil
and protein

Stages in germination: epigeal

first true leaves
are hairy

cotyledons brought
above ground and
become green and
photosynthetic

cotyledons
are not hairy

apical meristem

cotyledons being
withdrawn from
pericarp

pericarp may
still be attached
to cotyledons, or
may remain below
ground

cotyledons
open out

pericarp
splits

plumule hook
protecting
plumule

lateral
root

tap
root

root hairs

radicle emerges
and develops into
tap root

A **cypsela** is a one-seeded fruit made from an inferior ovary. It is found in the
family Compositae, and the pericarp and testa are separate. Often a pappus of hairs
is attached to the fruit, e.g. dandelion.
Sunflowers are native to Mexico and western North America.
The food store is mainly oil and protein; although the cotyledons store endosperm,
they are thin and not fleshy because oil is a very efficient food store of high calorific
value, so less is needed than of a carbohydrate food store.

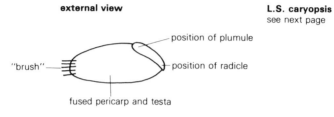

Triticum species	Division	Spermatophyta
wheat	Subdivision	Angiosperms
	Class	Monocotyledons
	Family	Gramineae
seed and germination	Genus	Triticum

Caryopsis (fruit) of wheat: endospermic seed

external view

L.S. caryopsis
see next page

position of plumule

"brush"

position of radicle

fused pericarp and testa

No micropyle is visible because this is a fruit, whose outer layer is pericarp.

Stages in germination: hypogeal

L.S. coleoptile

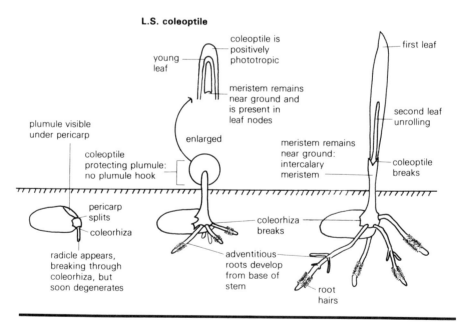

A **caryopsis** is a one-seeded fruit in which the testa and pericarp are fused, and is found in the family Gramineae (grasses).

Wheat is native to the Old World and has been cultivated since Neolithic times. The food store is starch and protein. The protein is highly deficient in the amino acid lysine. This is plentiful in the pulses (legumes), and a mixture of pulses and cereals provides most of the acids essential to man, although it may be slightly deficient in methionine, which is very low in legumes and slightly deficient in cereals.

Germination is hypogeal, because the scutellum remains below ground.

Division	Spermatophyta
Subdivision	Angiosperms
Class	Monocotyledons
Family	Gramineae
Genus	Triticum

L.S. caryopsis (fruit) of wheat, cut through embryo

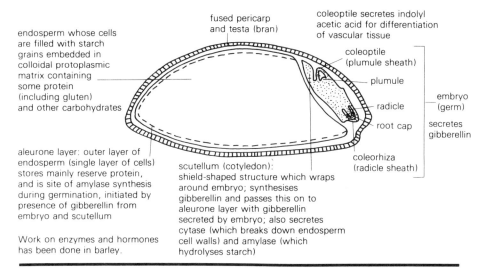

endosperm whose cells are filled with starch grains embedded in colloidal protoplasmic matrix containing some protein (including gluten) and other carbohydrates

fused pericarp and testa (bran)

coleoptile secretes indolyl acetic acid for differentiation of vascular tissue

coleoptile (plumule sheath)

plumule

radicle

embryo (germ)

root cap — secretes gibberellin

aleurone layer: outer layer of endosperm (single layer of cells) stores mainly reserve protein, and is site of amylase synthesis during germination, initiated by presence of gibberellin from embryo and scutellum

scutellum (cotyledon): shield-shaped structure which wraps around embryo; synthesises gibberellin and passes this on to aleurone layer with gibberellin secreted by embryo; also secretes cytase (which breaks down endosperm cell walls) and amylase (which hydrolyses starch)

coleorhiza (radicle sheath)

Work on enzymes and hormones has been done in barley.

The wheat caryopsis (i.e. grain) is used in bread in the following ways:
White bread contains endosperm ony, but the B vitamins thiamine and nicotinic acid, and the minerals iron and calcium are added; this is required by law.
Brown bread is made from brown flour, which is a mixture of white and wholemeal, or white flour with bran added, or white flour coloured with caramel.
Wheatgerm bread is made of brown flour with wheat germ (i.e. embryo) added.
Wholemeal or wholewheat bread contains the whole grain, including the germ and bran.
Granary bread is made from brown flour with added malted flour and the kernels of wheat and other cereals.

The endosperm contains starch and protein including gluten which is necessary to make bread rise.

The embryo (germ) contains most vitamins of the B complex (except B_{12}) and vitamin E, the minerals calcium, magnesium, potassium, phosphorus and iron, and some unsaturated fatty acids.

The coat (bran) contains fibre.

Notes: rice is very similar, but has an inedible out husk made of persistent lemma and pale (see grass flower, page 121). White and polished rice consist of the endosperm only, and diets based on it are deficient in the B complex vitamins, especially thiamine, resulting in the deficiency disease beri-beri. Brown rice contains the whole grain with outer husk removed. In rice, the pericarp and testa are more separate; the pericarp forms the outer brown bran layer, and the testa forms the inner creamy white one.

In cereals, the vitamin nicotinic acid (also called niacin or vitamin PP but considered part of the B complex) is in a bound form called niacytin, but usually the vitamin deficiency disease pellagra does not occur, because it can be made from tryptophan. But a diet of exclusively maize, which lacks tryptophan, can lead to pellagra.

Zea mays maize seed and germination	Division	Spermatophyta
	Subdivision	Angiosperms
	Class	Monocotyledons
	Family	Gramineae
	Genus	Zea

Caryopsis (fruit) of maize: endospermic seed

external view

**L.S. caryopsis
cut through embryo**

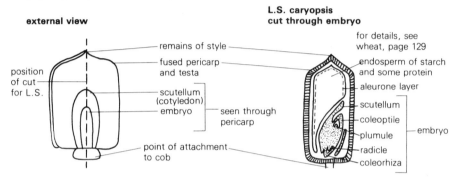

position of cut for L.S.

remains of style
fused pericarp and testa
scutellum (cotyledon)
embryo — seen through pericarp
point of attachment to cob

for details, see wheat, page 129
endosperm of starch and some protein
aleurone layer
scutellum
coleoptile
plumule
radicle
coleorhiza
embryo

No micropyle is visible because this is a fruit, whose outer layer is pericarp.

Stages in germination: hypogeal

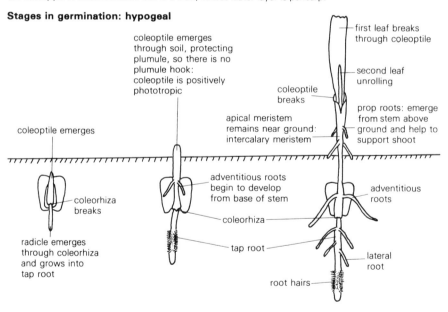

coleoptile emerges through soil, protecting plumule, so there is no plumule hook: coleoptile is positively phototropic

first leaf breaks through coleoptile

second leaf unrolling

coleoptile breaks

prop roots: emerge from stem above ground and help to support shoot

coleoptile emerges

apical meristem remains near ground: intercalary meristem

coleorhiza breaks

adventitious roots begin to develop from base of stem

coleorhiza

tap root

adventitious roots

radicle emerges through coleorhiza and grows into tap root

lateral root

root hairs

All cereal grains are fairly similar; for details see wheat, pages 128–9.

Maize is native to the New World. The food store is different from wheat in that it stores a lot of oil in the embryo, as well as starch and protein in the endosperm. Besides lacking lysine, its protein is particularly deficient in the amino acid tryptophan, which is the precursor of the vitamin nicotinic acid (niacin). Diets relying exclusively on maize can show the vitamin deficiency disease, pellagra.

Sweet corn is a variety of maize containing sugar and more water.

Vegetative propagation and perennation in angiosperms

Vegetative propagation (or reproduction) is any method of non-sexual reproduction involving the production of large, multicellular masses.

Perennation is any method of passing the winter. Angiosperms perennate in three ways:

1 By seeds. This is the only method in annual plants where the entire plant dies down in winter, and only the seed survives until next spring.

2 Woody perennials, i.e. trees or shrubs. These have a woody trunk and stand the winter above ground. They often shed their leaves to reduce the need for water, which may be frozen as ice or snow.

3 Herbaceous perennials. These die down above ground in winter and survive underground, usually with a food store in some storage organ. These storage organs are often also organs of vegetative reproduction. They are often used commercially for vegetative propagation as they produce plants identical to the parent. In general, sugars made in photosynthesis are passed via the phloem to the storage organ where they are usually condensed to insoluble polysaccharides. In spring, these polysaccharides are hydrolysed to soluble disaccharides or monosaccharides and pass to buds in storage organs where they are used for growth.

Note: biennials such as carrot, have an underground storage organ in the first winter, but the next winter survive only as the seeds.

Any part of the plant may be used as a perennating organ and organ of vegetative propagation. Definitions of the terms below are found in the glossary.

Stems may be:
(a) Stem tubers, e.g. potato, for perennation and vegetative propagation.
(b) Rhizomes, e.g. iris, ginger, for perennation and vegetative propagation.
(c) Rootstocks, e.g. chrysanthemum, for perennation and vegetative propagation.
(d) Corms, e.g. crocus, gladiolus, for perennation and vegetative propagation.
(e) Runners, e.g. strawberry, for vegetative propagation only.
(f) Stolons, e.g. couch grass, for vegetative propagation only.
(g) Suckers, e.g. roses, bananas, for vegetative propagation only.
(h) Offsets (small runners) e.g. houseleeks, for vegetative propagation only.

Leaves may be:
(a) Bulbs, e.g. onion, daffodil, for perennation and vegetative propagation.
(b) Occasionally young plants may grow from the edges of leaves, either naturally, e.g. walking fern, or by leaf cuttings, e.g. *Peperomia*, some begonias.

Roots may be:
(a) Root tubers, e.g. dahlia, lesser celandine, for perennation.
(b) Swollen tap root, e.g. carrot, for overwintering of a biennal. Roots form buds less readily than stems, so root tubers usually must have a little piece of stem, or stem buds, attached to them in order to grow in spring.

Perennation may also occur by detached winter buds called turions, especially in water plants. Vegetative propagation may occur by little bulbils which can replace flowers, e.g. some onions.

Solanum tuberosum	Division	Spermatophyta
	Subdivision	Angiosperms
potato	Class	Dicotyledons
	Family	Solanaceae
stem tuber	Genus	Solanum

1 Potato in the ground contains starch food store in parenchyma cells. The corky impermeable epidermis contains protein (mainly enzymes) under the surface.

Growth will occur from buds, "eyes".

2 In spring, the tuber starts to grow. Enzymes become active. Amylase digests starch to maltose which passes to buds. These begin to form shoots and grow into new plants, using food stored in the tuber.

Sometimes two or more buds start to develop, but in nature only one will form a plant.

3 Shoot puts down adventitious roots and grows into a new plant. It begins to photosynthesise, and uses its roots to absorb water and mineral salts. It no longer uses food from the tuber, which is now shrunken.

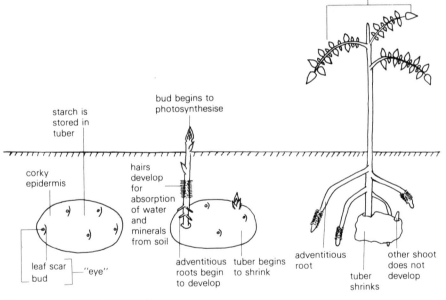

photosynthesising plant developed from bud

bud begins to photosynthesise

starch is stored in tuber

corky epidermis

hairs develop for absorption of water and minerals from soil

leaf scar
bud — "eye"

adventitious roots begin to develop

tuber begins to shrink

adventitious root

tuber shrinks

other shoot does not develop

Diagrams are not to the same scale.

A potato is a **stem tuber** which is a swollen underground end part of a stem. It can be recognised as a stem by the leaf scars and buds, the "eyes".

The food store is mainly starch, with protein immediatly under the skin, and good quantities of vitamin C (which declines in storage) with some B vitamins and minerals. The skin contains fibre.

Potatoes are native to the temperate Andes of South America, and were brought to Europe in the 16th century.

4 In summer, potato begins to make new tubers underground, and sugars from photosynthesis are passed down to them in the phloem. In the tubers, sugars are converted to starch and stored in plastids in parenchyma cells.

5 In winter, the old plant dies down and tubers are left in the soil over winter. Next year, each potato will develop into a new tuber.

Seeds also lie in the ground and germinate into new plants in spring.

Tubers are used for both vegetative reproduction and perennation.

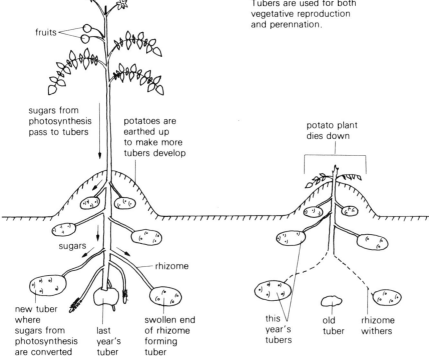

Vegetative reproduction from ''seed'' potatoes produces plants identical to the parents, so that constant varieties can be produced, e.g. King Edward, Maris Piper. Sexual reproduction by seeds produces plants which are not identical to the parents, so commercial potatoes are not propagated from seed.

Earthing up increases the number of potatoes, as they are only formed underground. In the light, potatoes produce chlorophyll, and also a poison, solanine, which is why green potatoes (and green parts of the plant) are poisonous.

Iris germanica	Division	Spermatophyta
iris	Subdivision	Angiosperms
	Class	Monocotyledons
	Family	Iridaceae
rhizome	Genus	*Iris*

1 In winter, the rhizome in the ground contains starch and other polysaccharides in parenchyma cells. The thick epidermis has scars of last year's foliage leaves.

2 In spring, growth in length occurs by the terminal bud which curves upwards and produces leaves, flowers, and a horizontal stem. A lateral bud may also develop. Food stored in the rhizome is used for growth until photosynthesis begins. As a result, the rhizome shrinks.

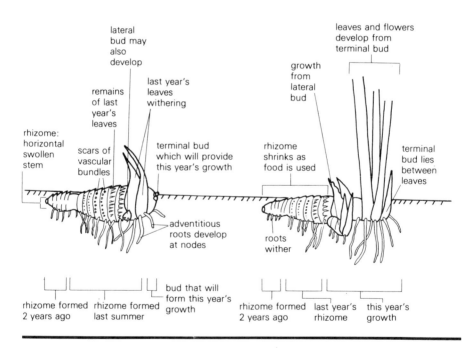

A **rhizome** is a horizontal, typically underground stem, usually but not always swollen. When swollen, as in iris, it is an organ of vegetative reproduction and perennation. When not swollen, as in couch grass, it is an organ of vegetative reproduction only.

The food store is mainly starch.

The genus *Iris* is native to the northern hemisphere, and garden bearded irises are mostly hybrids with a complex parentage, and must be propagated vegatively.

''Root'' ginger is a rhizome which contains aromatic oils as well as starch, giving the spicy taste.

3   Sugars made in photosynthesis are passed via the phloem to the rhizome and stored as starch. A new rhizome is formed at the front of the plant. The lateral bud also forms a new rhizome. Seed and fruit are also produced.

4   In winter, leaves die down and rhizomes pass the winter underground. Eventually the old connection between branches of the rhizome wither so that there are 2 plants where there was one before. This is vegetative propagation as well as perennation.

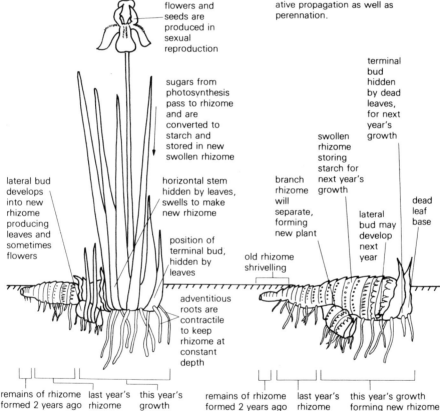

flowers and seeds are produced in sexual reproduction

sugars from photosynthesis pass to rhizome and are converted to starch and stored in new swollen rhizome

lateral bud develops into new rhizome producing leaves and sometimes flowers

horizontal stem hidden by leaves, swells to make new rhizome

position of terminal bud, hidden by leaves

adventitious roots are contractile to keep rhizome at constant depth

terminal bud hidden by dead leaves, for next year's growth

swollen rhizome storing starch for next year's growth

branch rhizome will separate, forming new plant

old rhizome shrivelling

lateral bud may develop next year

dead leaf base

remains of rhizome formed 2 years ago last year's rhizome this year's growth

remains of rhizome formed 2 years ago last year's rhizome this year's growth forming new rhizome

Vegetative reproduction produces plants identical to the parent, while sexual reproduction forms seeds which produce plants showing variation.

The importance of vegetative reproduction is that it enables the plant, once established, to spread over a wide area of the habitat. This is particularly important in pteridophytes, where the sporophyte can reproduce by rhizomes without having to go through the vulnerable gametophyte stage, thus enabling horsetails and bracken to grow in dry places.

The iris rhizome is unusual in being partially on the surface.

135

Crocus species		
crocus	*Division*	Spermatophyta
	Subdivision	Angiosperms
	Class	Monotyledons
	Family	Iridaceae
corm	*Genus*	*Crocus*

1 Corm stores starch in parenchyma cells, and is protected by brown papery scales leaves.

Growth occurs by the main bud, and sometimes also by a lateral bud.

2 Bud develops into new corm. Starch is converted to sugars and passes to bud as it grows. The corm becomes smaller. Leaves begin to photosynthesise.

2a Only one bud develops.   2b Sometimes two buds develop.

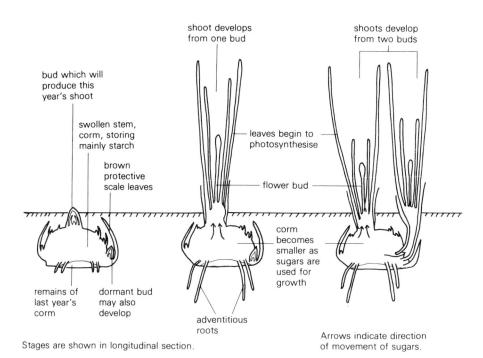

bud which will produce this year's shoot

swollen stem, corm, storing mainly starch

brown protective scale leaves

shoot develops from one bud

leaves begin to photosynthesise

flower bud

shoots develop from two buds

corm becomes smaller as sugars are used for growth

remains of last year's corm

dormant bud may also develop

adventitious roots

Stages are shown in longitudinal section.

Arrows indicate direction of movement of sugars.

A **corm** is a swollen underground upright stem.
The food store is mainly starch.
    The genus *Crocus* is native to temperate Europe, especially the Mediterranean area with hot summers and cool, moist springs and winters.
    Corms are also found in gladioli, cuckoo pint, and bulbous buttercup.

3 Crocus produces flowers and seeds. Sugars formed in photosynthesis pass into phloem and are transported to stem base which swells up to form new corm. Sugars are converted to starch and stored in the corm. Contractile roots on the new corm pull it below ground.

4 New corms enlarge with food made by photosynthesis. The old plants die down. Next year each corm will develop into a new crocus plant.

3a If one bud develops, one corm is formed.

3b If two buds develop, two corms are formed.

4a If only one corm is produced, this is perennation, but not reproduction.

4b If two corms are produced, this is reproduction as well as perennation.

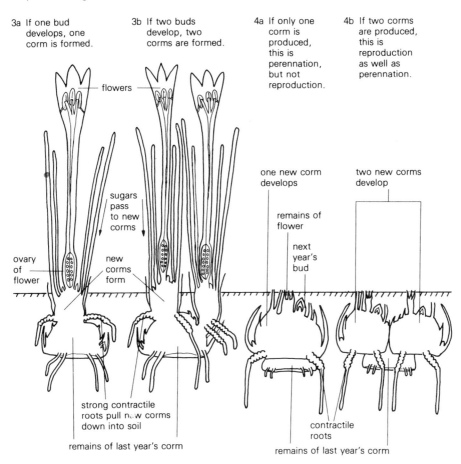

flowers

sugars pass to new corms

one new corm develops

two new corms develop

remains of flower

next year's bud

ovary of flower

new corms form

strong contractile roots pull new corms down into soil

remains of last year's corm

contractile roots

remains of last year's corm

Vegetative reproduction produces plants identical to the parent while sexual reproduction by seed produces crocuses showing variation. Crocuses are sometimes grown from seed, but corms are preferred as they may be planted in autumn and produce flowers next spring. If grown from seed, the seed must be planted in spring, and does not produce flowers until the following year.

***Narcissus* species** daffodil **bulb**	**Division** Spermatophyta
	Subdivision Angiosperms
	Class Monocotyledons
	Family Amaryllidaceae
	Genus *Narcissus*

1 Bulbs store food in leaf bases, often as sugars rather than starch. Bulb is protected by papery scale leaves.

Growth occurs by the main bud, and sometimes also by a lateral bud.

2 Bud develops into a new shoot, taking food from leaf bases, so that the bulb shrinks. Then the leaves start to photosynthesise.

2a Only main bud develops into a shoot.

2b Sometimes main and lateral buds develop into shoots.

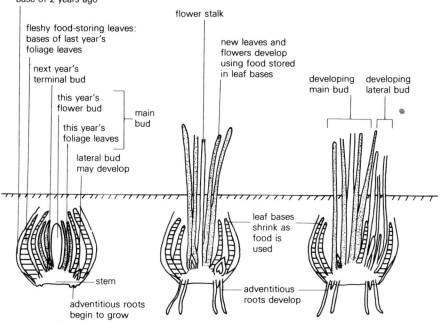

outer protective scale leaf: remains of leaf base of 2 years ago

fleshy food-storing leaves: bases of last year's foliage leaves

next year's terminal bud

this year's flower bud

this year's foliage leaves

lateral bud may develop

main bud

stem

adventitious roots begin to grow

flower stalk

new leaves and flowers develop using food stored in leaf bases

leaf bases shrink as food is used

adventitious roots develop

developing main bud

developing lateral bud

Bulbs are shown in longitudinal section.
Bulbs are simplified, not all scale leaves are shown.

Diagrams are not to the same scale.

A **bulb** stores food in underground swollen modified leaves, which may be the leaf bases of foliage leaves, or specially produced scale leaves.

The food store is sugars, which cause water to be taken up by osmosis, making the bulbs crisp.

Daffodils are native to Europe, W. Asia and N. Africa; most of our garden forms are hybrids.

Other plants with bulbs include bluebell, snowdrop, hyacinth, tulip, onion.

3 Plants produce flowers and seeds. Sugars made in photosynthesis are passed down leaves to leaf bases which swell to form a new bulb.

4 New bulb enlarges with sugars made in photosynthesis. Old plants die down. Next year each bulb will develop into a new plant.

3a If one bud developed, one bulb forms.

3b If two buds developed, two bulbs form.

4a If only one bulb is formed, this is perennation, not reproduction.

4b If two bulbs are formed, this is reproduction as well as perennation.

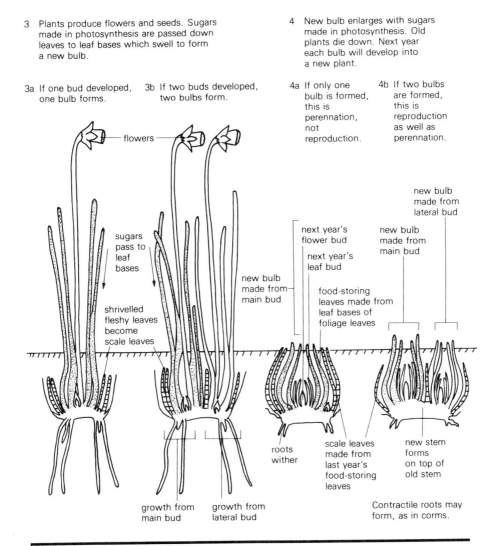

Vegetative reproduction produces plants identical to the parents, so that varieties can be kept constant, e.g. King Alfred daffodils, black tulips, ''Ailsa Craig'' onions. Seeds are also produced in most of these plants, and they can also be grown from seed (as onions often are).

All of the onion group are modified bulbs; the onion is similar to the daffodil bulb; leeks are made of swollen leaf bases; in garlic the bulb is made of a number of miniature bulbs called cloves enclosed in the skin of the parent bulb.

Fragaria x ananassa	Division	Spermatophyta
strawberry	Subdivision	Angiosperms
	Class	Dicotyledons
runner: vegetative	Family	Rosaceae
propagation	Genus	*Fragaria*

Parent plant with runner

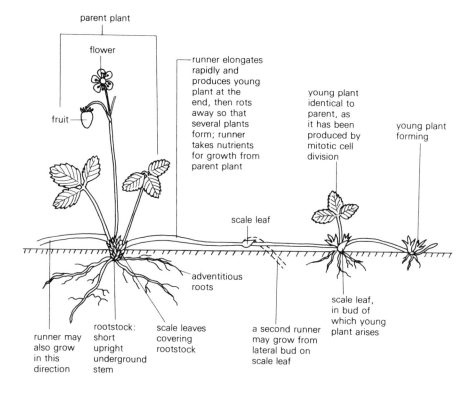

For detail of flower and fruit, see page 106.

parent plant

flower

fruit

runner elongates rapidly and produces young plant at the end, then rots away so that several plants form; runner takes nutrients for growth from parent plant

young plant identical to parent, as it has been produced by mitotic cell division

young plant forming

scale leaf

adventitious roots

scale leaf, in bud of which young plant arises

runner may also grow in this direction

rootstock: short upright underground stem

scale leaves covering rootstock

a second runner may grow from lateral bud on scale leaf

A **runner** is a horizontal thin stem running along the surface of the soil, and producing a new plant at the end.

Runners are for vegetative reproduction, not perennation.

Strawberries survive the winter as the whole plant above soil and may be protected from frost by straw.

Our modern strawberries are hybrids and are grown from plants produced from runners, but alpine strawberries may be grown from seed.

Runners are also found in the house-plant, *Chlorophytum*, the spider plant.

Notes: a **stolon** is similar to a runner, but young plants are produced all along the length, not just at the end, because internodes are shorter.

Division	Spermatophyta
Subdivision	Angiosperms
Class	Dicotyledons
Family	Ranunculaceae
Genus	*Ranunculus*

Ranunculus ficaria
lesser celandine

root tuber: perennation

Entire plant

leaves arise from a short stem,
i.e. apparently from root and
are called radicle leaves

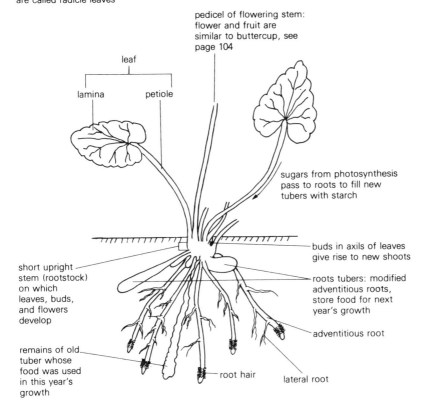

pedicel of flowering stem:
flower and fruit are
similar to buttercup, see
page 104

leaf

lamina   petiole

sugars from photosynthesis
pass to roots to fill new
tubers with starch

buds in axils of leaves
give rise to new shoots

short upright
stem (rootstock)
on which
leaves, buds,
and flowers
develop

roots tubers: modified
adventitious roots,
store food for next
year's growth

adventitious root

remains of old
tuber whose
food was used
in this year's
growth

root hair   lateral root

A **root tuber** is a swollen food-storing adventitious root.
Root tubers are mainly organs of perennation only, because buds usually develop
on stems but not on roots, although this may happen occasionally, e.g. phlox.
Propagation by root tuber can usually only be done if a piece of stem and bud is
included with the root.
    The food store in root tubers is usually polysaccharide but not always starch, e.g.
dahlias store inulin, made of fructose units. Sweet potatoes are root tubers which
contain some sugars and protein as well as starch.
    Lesser celandine is an early-flowering buttercup of woods, grassy banks and
meadows, the food stored in its tubers enabling it to flower in early spring.

Daucus carota	Division	Spermatophyta
carrot	Subdivision	Angiosperms
	Class	Dicotyledons
	Family	Umbelliferae
swollen tap root	Genus	*Daucus*

First spring
Carrots can only be grown from seed, not from carrots.

The seed germinates, producing leaves and a tap root.

First summer
Photosynthesis produces sugars which are passed in the phloem to the tap root. The root swells up as parenchyma cells store sugars.

No flowers are produced this year.

First winter
Upper parts die down. The carrot is left in the soil.

Carrots are usually harvested at this stage.

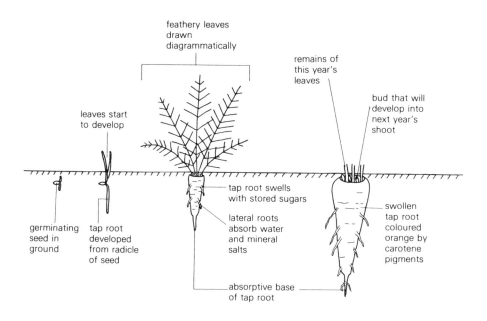

feathery leaves drawn diagrammatically

remains of this year's leaves

bud that will develop into next year's shoot

leaves start to develop

germinating seed in ground

tap root developed from radicle of seed

tap root swells with stored sugars

lateral roots absorb water and mineral salts

absorptive base of tap root

swollen tap root coloured orange by carotene pigments

A **biennial** is a plant that completes its life cycle in two years. The carrot must be grown from seed; in its first year it stores food in its swollen tap root, and overwinters in the ground in this form. In its second year, the food in the carrot is used to produce leaves, then flowers and fruits are formed, using sugars made in photosynthesis. In the second year no carrot is formed, and reproduction is only by seeds.

Carrots are not perennials since the swollen root is not produced every year, and one plant does not survive from year to year.

Life cycle of a biennial

Second spring
A bud on the carrot develops into a new shoot, using sugars stored in the carrot, so carrot shrinks.

Second summer
Flowers and fruits are produced and seeds are dispersed. This year, no swollen root is produced.

Second winter
The whole plant dies down. Only the seed is left in the soil.

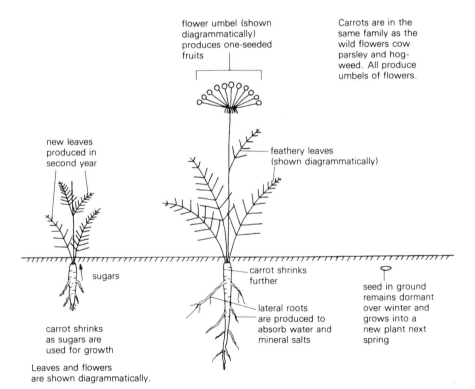

new leaves produced in second year

flower umbel (shown diagrammatically) produces one-seeded fruits

feathery leaves (shown diagrammatically)

Carrots are in the same family as the wild flowers cow parsley and hog-weed. All produce umbels of flowers.

sugars

carrot shrinks as sugars are used for growth

carrot shrinks further

lateral roots are produced to absorb water and mineral salts

seed in ground remains dormant over winter and grows into a new plant next spring

Leaves and flowers are shown diagrammatically.

The food store is mainly sugars. The orange colour is due to carotene, which can be converted in the body to vitamin A.

Carrots are probably native to Afghanistan and have been cultivated since Greek times. At first they were white, or yellow due to carotenoids, or purple due to anthocyanins. Purple types were not popular because they cooked to a bad colour. The orange form was developed in Holland in the 15th century.

Parsnips, beetroots and the beets are also biennials and complete their life cycles in a similar way.

Advantages of perennation and vegetative propagation

Advantages of perennation

1 The food stored in a perennating organ enables a plant to grow rapidly in early spring, so that it can flower and reproduce early. This occurs in many spring-flowering bulbs and corms such as daffodil, snowdrop, bluebell. It is particularly important in woodland plants, because they can photosynthesise, produce seeds, and refill their perennating organs before the tree canopy forms and reduces the light reaching the woodland floor.

2 The plant can grow in the same place from year to year. Many seeds do not find a suitable site for germination, but once established in a habitat, the plant is more sure of survival if it perennates. However, seeds are necessary too, because the plant must produce bodies which are dispersed over long distances to colonise new habitats in order to increase the population of the plant in many areas.

3 Perennation is less useful to plants in disturbed habitats where they may be destroyed annually, such as cornfields, or areas covered by glaciers at times. This is why some weeds are annuals. But many plants that we cultivate as annuals are, in fact, perennials from warmer climates which are destroyed by frost in winter.

4 Man has selected perennating organs and increased their size for his own food supply, e.g. potato, onion, etc.

Advantages of vegetative propagation

1 A continuous food supply is available from the parent to the progeny until it is well established.

2 If the parent is growing in a favourable environment, the progeny which are produced by mitosis and so identical to the parent, will also be adapted to the environment and able to take full benefit from it.

3 Once the parent is established in a habitat, it is easier for the progeny to be established nearby and to select the best place for growth. For example, plants growing from a runner can choose the position at which a new plant is formed, avoiding other species or poor ground, in a way that seeds cannot. The parent plant may be large and strong, which can help the progeny to compete with other established plants.

4 Some weeds produce dense and very persistent clumps by vegetative propagation, using rhizomes, rootstocks or bulbils. If a gardener tries to pull up the weed, only a little of the propagating structure needs to survive in the soil for the plant to persist, making such weeds very difficult to eradicate.

5 Vegetative propagation is useful to the horticultural grower, because he knows that he will produce pure lines of plants identical to his parent stock. But there may be problems, such as that all plants will be equally susceptible to disease.

Vegetative propagation alone is not enough. It does not enable the progeny to become dispersed over long distances, and although being identical to the parent is a short term advantage, in the long term variation is necessary, so that the plants can evolve and adapt to environmental changes. Variation is best brought about by sexual reproduction with the production of seeds.

Alternation of generations and its significance

Alternation of generations means that in the life cycle of a plant there are two types of plant body, i.e. two different generations; these are a haploid gametophyte generation which produces gametes by mitosis, and a diploid sporophyte generation which produces spores by meiosis. Each generation gives rise to the other so that the generations alternate in the life cycle.

Alternation of generations is found in some algae, and in all bryophytes, pteridophytes and spermatophytes. The diagram below is a generalised diagram of alternation of generations and applies to all these groups.

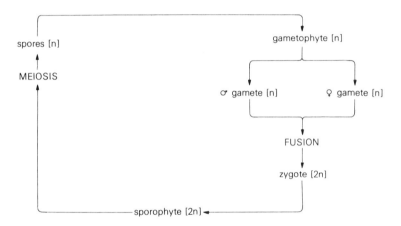

Generalised diagram of alternation of generations

In bryophytes, pteridophytes and spermatophytes an evolutionary sequence can be seen in the nature of the two generations, and there is a gradually increasing dominance of the sporophyte generation with evolutionary advance.

In bryophytes, the gametophyte is the independent and most conspicuous generation, while the sporophyte, although when present very conspicuous, is dependent on the gametophyte. The gametophyte is always low-growing and is the leafy plant in mosses and the thallus or leafy plant in liverworts. It is usually not well adapted to life on land, since many bryophytes have no cuticle to prevent drying out, although this may be present in mosses which live in dry places. The gametophyte also has no vascular tissue for conduction and support, and has only rhizoids, not roots, for absorption of water; rhizoids penetrate only the surface layers, so once these have dried out the plant cannot absorb water from depth in the soil. Water is also needed for sexual reproduction, since the male gamete is a flagellated spermatozoid which swims to the female gamete, so bryophytes usually live in damp places even though some mosses can withstand drying out. The gametophyte cannot grow tall because it has no vascular tissue and because of the swimming sperm. The sporophyte in bryophytes consists of the seta and capsule, which sticks up into a slightly drier microclimate and requires dry conditions for spore dispersal.

145

In pteridophytes, the sporophyte is the dominant conspicuous generation, but it is dependent on the gametophyte at the start of its existence. Pteridophytes have overcome some of the problems of bryophytes. In this group, the sporophyte is the fern, horsetail or club moss plant. It has a cuticle on the leaves and stems to conserve water, and stomata for the entry and exit of gases. There is xylem to transport water and to provide strength, and true roots to penetrate deep into the soil for water. So the sporophyte is well adapted to life on land.

But the spores produced by the sporophyte grow into the independent photosynthetic gametophyte generation called the prothallus, which is similar to a reduced liverwort thallus and is equally vulnerable to drying out. The male gamete, as in bryophytes, is a swimming spermatozoid, so water is necessary for both the survival of the gametophyte and for sexual reproduction. After fertilisation, the sporophyte begins to develop on the gametophyte for a while before it takes up an independent existence. This is why most pteridophytes live in damp places — because of the demands of the gametophyte on which the sporophyte originally depends. Heterosporous pteridophytes like *Selaginella* have partly overcome the problem, since the gametophyte is more dependent on the sporophyte and is mainly nourished by it.

There are some pteridophytes which inhabit dry environments, for example bracken, field horsetail, and some "resurrection plant" *Selaginellas*. These plants survive by spreading vegetatively by their rhizomes, and in the horsetail by tubers as well, and avoiding sexual reproduction and the vulnerable gametophyte stage.

Pteridophytes have been likened in evolutionary terms to amphibians, in that both are in a half-way stage to life on land, with one stage quite well adapted to land, the sporophyte and adult, but both require water for sexual reproduction and for the other stage of the life cycle, the gametophyte and tadpole. Pteridophytes and amphibians were the dominant flora and fauna of the Carboniferous period, when the wet, swampy conditions were ideal for their modes of life. It was the drier conditions of the following period, the Permian, which led to the evolution of truly land-adapted organisms, the spermatophytes and reptiles.

In spermatophytes, the sporophyte is the dominant and independent generation, and is even better adapted to life on land than pteridophytes, while the gametophyte is very reduced and dependent on the sporophyte. Spermatophytes have overcome the problems of pteridophytes, i.e. the free, independent gametophyte stage and swimming spermatozoids. The trend is beginning in heterosporous pteridophytes, where the gametophyte is becoming more dependent for food on the sporophyte, but all pteridophytes require free water for sexual reproduction. Spermatophytes have overcome this difficulty by retaining the gametophyte generation on the sporophyte. In this group, the sporophyte is the main plant, the cycad, conifer or flowering plant. There are always two kinds of spore, microspores and megaspores, which give rise to separate male and female gametophytes. The female gametophyte is retained on its parent sporophyte, and the male gametophyte is brought to it inside its microspore (pollen grain) in the process of pollination. The male gamete does not swim in free water to the female gamete but is taken to it in the pollen tube, which grows to the female gamete.

In bryophytes and pteridophytes, the male gamete is attracted to the female gamete by a chemical and reaches it by a swimming movement (positive chemotaxis). In spermatophytes, it used to be thought that the ovule attracted the pollen tube by a chemical and that it reached the ovule by a growth movement (positive chemotropism), but it is now believed that the pollen tube is merely growing away from the oxygen in air (negative aerotropism).

Cycads are very significant in that they have pollination and a pollen tube like other spermatophytes, but the spermatozoid has flagella and swims inside the ovule to reach the female gamete. There is no physical "need" for swimming sperm, and it is thought to be a vestigial characteristic, indicating that spermatophytes evolved from pteridophyte ancestors with swimming male gametes.

Spore dispersal
In all three divisions, although wet conditions are required for sexual reproduction, dry conditions are needed for spore dispersal, resulting in hygroscopic movement of the part of the structure containing the spores. In mosses, the movement is of the peristome teeth, in liverworts and horsetails of elaters, and in ferns of the annulus. Microsporangia and megasporangia open by longitudinal slits in dry weather, and anthers curl back by movement of fibrous cells. In all cases, spores are dispersed by air currents or wind, but in angiosperms insects also used for the transmission of microspores in pollination.

The botanical nature of the edible parts of our food plants

Common name	Latin name	Family	Botanical nature of part eaten
almond	*Prunus dulcis*	Rosaceae (D)	drupe; shell of nut is endocarp and edible part is seed
apple	*Malus* hybrids	Rosaceae (D)	pome; edible part is receptacle
apricot	*Prunus armeniaca*	Rosaceae (D)	drupe
artichoke, globe	*Cynara scolymus*	Compositae (D)	bracts of inflorescence
artichoke, Jerusalem	*Helianthus tuberosus*	Compositae (D)	stem tuber
asparagus	*Asparagus officinalis*	Liliaceae (M)	young shoot growing from rootstock
aubergine	*Solanum melongena*	Solanaceae (D)	berry
avocado	*Persea americana*	Lauraceae (D)	drupe with much fat and protein
banana	*Musa*	Musaceae (M)	seedless berry
barley	*Hordeum* species	Gramineae (M)	fruit, caryopsis
bean, broad	*Vicia faba*	Leguminosae (D)	seed
bean, French	*Phaseolus vulgaris*	Leguminosae (D)	} immature seed and pod
bean, runner	*Phaseolus coccineus*	Leguminosae (D)	
bean, mung (beansprouts)	*Vigna aureus*	Leguminosae (D)	seedling with cotyledons
beetroot	*Beta vulgaris*	Chenopodiaceae (D)	swollen tap root: colour — anthocyanins
blackberry	*Rubus ulmifolius*	Rosaceae (D)	drupelets
black currant	*Ribes nigrum*	Grossulariaceae (D)	berry
Brazil nut	*Bertholletia excelsa*	Lecythidaceae (D)	seed; woody testa is the nut's shell
broccoli	*Brassica oleracea*	Cruciferae (D)	loose flower head with fertile flowers
Brussels sprout	*Brassica oleracea*	Cruciferae (D)	vegetative buds
cabbage	*Brassica oleracea*	Cruciferae (D)	rosette of foliage leaves
caraway	*Carum carvi*	Umbelliferae (D)	dry fruit
carrot	*Daucus carota*	Umbelliferae (D)	swollen tap root of biennial
cashew	*Anacardium occidentale*	Anacardiaceae (D)	seed, which is kernel of true nut hanging from ''cashew apple'' of enlarged stalk
cauliflower	*Brassica oleracea*	Cruciferae (D)	flower head of very tightly packed sterile flower buds

D = dicotyledon; M = monocotyledon

Common name	Latin name	Family	Botanical nature of part eaten
celery	*Apium graveolens*	Umbelliferae (D)	petioles of leaves growing from reduced stem
cherry	*Prunus* species	Rosaceae (D)	drupe
chestnut (sweet)	*Castanea sativa*	Fagaceae (D)	true nut with spiny cupule which is removed before sale
chicory	*Cichorium intybus*	Compositae (D)	root is used as coffee substitute; leaves in salad, usually blanched
cinnamon	*Cinnamomum zeylanicum*	Lauraceae (D)	strips of dried bark, called quills
clove	*Eugenia caryophyllus*	Myrtaceae (D)	dried flower bud
cob nut	another name for hazel		
cocoa	*Theobroma cacao*	Sterculiaceae (D)	seed
coconut	*Cocos nucifera*	Palmae (M)	drupe
coffee	*Coffea* species	Rubiaceae (D)	seeds "beans" inside berry
coriander	*Coriandrum sativum*	Umbelliferae (D)	dry fruit; leaves
corn, sweet	*Zea mays*	Gramineae (M)	fruit, caryopsis
cress	*Lepidium sativum*	Cruciferae (D)	seedling
cucumber	*Cucumis sativus*	Cucurbitaceae (D)	specialised berry, a pepo
cumin	*Cuminum cyminum*	Umbelliferae (D)	dry fruit
date	*Phoenix dactylifera*	Palmae (M)	one-seeded berry; stone is testa
endive	*Cichorium endivia*	Compositae (D)	leaves; often blanched
fig	*Ficus carica*	Moraceae (D)	multiple fruit made from whole inflorescence
filbert	*Corylus maxima*	Betulaceae or Corylaceae (D)	nut, like hazel but with a longer cupule
garlic	*Allium sativum*	Liliaceae (M)	small bulbs (cloves) surrounded by a common skin
ginger	*Zingiber officinale*	Zingiberaceae (M)	rhizome
grape	*Vitis vinifera*	Vitidaceae (D)	berry
currant sultana raisin	varieties of *Vitis vinifera*	Vitidaceae (D)	dried and rich in sugars
grapefruit	*Citrus paradisi*	Rutaceae (D)	berry (hesperidium)
gooseberry	*Ribes grossularia*	Grossulariaceae (D)	berry
hazelnut (cobnut)	*Corylus avellana*	Betulaceae or Corylaceae (D)	true nut

Common name	Latin name	Family	Botanical nature of part eaten
hop	*Humulus lupulus*	Cannabinaceae (D)	female flower
kohlrabi	*Brassica oleracea*	Cruciferae (D)	blanched swollen stem base
leek	*Allium ampeloprasum*	Liliaceae (M)	leaf bases of foliage leaves, forming elongated bulb
lemon	*Citrus limon*	Rutaceae (D)	berry (hesperidium)
lentil	*Lens culinaris*	Leguminosae (D)	seed
lettuce	*Lactuca sativa*	Compositae (D)	foliage leaves
mace	see nutmeg		
maize	see corn, sweet		
marrow	*Cucurbita pepo*	Cucurbitaceae (D)	pepo (as cucumber)
melon	*Cucumis melo*	Cucurbitaceae (D)	pepo (as cucumber)
mushroom	*Agaricus bisporus*	Agaricaceae (Fungi)	fruit body
mustard	*Sinapis alba*	Cruciferae (D)	seedling (or seed)
nutmeg	*Myristica fragrans*	Myristicaceae (D)	seed; mace is the aril of seed
oat	*Avena sativa*	Gramineae (M)	fruit, caryopsis
olive	*Olea europaea*	Oleaceae (D)	drupe
onion	*Allium cepa*	Liliaceae (M)	bulb
orange	*Citrus sinensis*	Rutaceae (D)	berry (hesperidium)
parsnip	*Pastinaca sativa*	Umbelliferae (D)	swollen tap root of biennial
pea	*Pisum sativum*	Leguminosae (D)	seed
peach	*Prunus persica*	Rosaceae (D)	drupe
peanut	*Arachis hypogaea*	Leguminosae (D)	seed; testa is thin brown coat, shell is pod (legume)
pear	*Pyrus communis*	Rosaceae (D)	pome
pecan	*Carya illinoensis*	Juglandaceae (D)	drupe (as walnut)
pepper, sweet	*Capsicum annuum*	Solanaceae (D)	berry
pepper, chilli	*Capsicum annuum*	Solanaceae (D)	berry
peppercorn	*Piper nigrum*	Piperaceae (D)	berry, removed before ripe, and dried
pineapple	*Ananas comosus*	Bromeliaceae (M)	multiple fruit made from whole inflorescence
pistachio	*Pistacia vera*	Anacardiaceae (D)	nut; endosperm forms the green kernel
plum	*Prunus* species	Rosaceae (D)	drupe

Common name	Latin name	Family	Botanical nature of part eaten
potato, Irish	*Solanum tuberosum*	Solanaceae (D)	stem tuber
potato, sweet	*Ipomoea batatas*	Convolvulaceae (D)	root tuber
quince	*Cydonia vulgaris*	Rosaceae (D)	pome
radish	*Raphanus sativus*	Cruciferae (D)	swollen tap root, colour due to anthocyanins
rhubarb	*Rheum rhaponticum*	Polygonaceae (D)	petiole of leaf growing from underground rootstock, colour due to anthocyanins; leaf is poisonous due to oxalic acid
rice	*Oryza sativa*	Gramineae (M)	fruit, caryopsis
rye	*Secale cereale*	Gramineae (M)	fruit, caryopsis
saffron	*Crocus sativus*	Iridaceae (M)	stigmas of flower
sesame	*Sesamum indicum*	Pedaliaceae (D)	seed
soya	*Glycine max*	Leguminosae (D)	seed
spinach	*Spinacea oleracea*	Chenopodiaceae (D)	foliage leaves
strawberry	*Fragaria x ananassa*	Rosaceae (D)	pseudocarp
sugar beet	*Beta vulgaris*	Chenopodiaceae (D)	swollen tap root
sugar cane	*Saccharum officinarum*	Gramineae (M)	swollen stem (ratoon)
sunflower ''seed''	*Helianthus annuus*	Compositae (D)	fruit, cypsela
swede	*Brassica napus*	Cruciferae (D)	swollen hypocotyl and epicotyl
tea	*Camellia sinensis*	Theaceae (D)	young leaves
tomato	*Lycopersicon esculentum*	Solanaceae (D)	berry
turnip	*Brassica rapa*	Cruciferae (D)	swollen hypocotyl
vanilla	*Vanilla planifolia*	Orchidaceae (M)	fruit, usually called a pod
walnut	*Juglans regia*	Juglandaceae (D)	drupe with inferior ovary
watercress	*Nasturtium officinale*	Cruciferae (D)	foliage leaves
wheat	*Triticum* species	Gramineae (M)	fruit, caryopsis

151

Glossary

achene: (1) broadly, a small, indehiscent, dry fruit usually produced from a single carpel and containing one seed; (2) particularly, such a fruit produced from a superior ovary, having the pericarp and testa free from one another; *cf.* caryopsis, cypsela.

actinomorphic: radially symmetrical; *cf.* zygomorphic.

adventitious root: a root arising from somewhere other than the radicle of the seed.

aerotropism: the growth movement of part of a plant in relation to the direction of oxygen in air.

akinete: a thick-walled spore in blue-green algae, which is formed from a vegetative cell and can survive unfavourable conditions, then germinates into a new filament or hormogonium.

aleurone layer: the outer layer of endosperm, containing protein, especially in a caryopsis.

alternation of generations: the condition of having two forms of plant body in a life cycle, a haploid gametophyte generation which produces gametes by mitosis, and a diploid sporophyte generation which produces spores by meiosis; each generation gives rise to the other so that the generations alternate in the life cycle.

amphigastria: the underleaves of a leafy liverwort; *sing.* amphigastrium.

anatropous: of ovules, bending downwards against the stalk.

androecium: the stamens of a flower, collectively.

anemophilous: having wind pollination; *n.* anemophily; *cf.* entomophilous.

anisogamy: the fusion of gametes which differ in size but not in shape; such gametes are called anisogametes; *a.* anisogamous; *cf.* isogamy, oogamy.

annual: a plant which completes its life cycle in one year and spends the winter only as a seed in the ground, the rest of the plant dying completely; *cf.* biennial, perennial.

annulus: (1) a ring of special cells with thickened walls concerned in the dehiscence of a fern sporangium; (2) a ring of large cells concerned in the dehiscence of the operculum of a moss capsule.

anther: the part of the stamen which produces the pollen grains.

antheridiophore: a stalk bearing antheridia or male receptacles.

antheridium: a male sex organ, which contains male gametes; *pl.* antheridia.

antherozoid: another name for spermatozoid.

antipodal cells: in the embryo sac of angiosperms, the three cells or nuclei furthest from the micropyle.

apex: the tip; *pl.* apices.

apical: at the tip.

apical cell: a cell at the tip of the stem, root, or thallus, which undergoes mitosis, resulting in growth.

apical meristem: in spermatophytes, a group of cells at the tip of the stem or root, which undergo mitosis, resulting in growth; only spermatophytes have apical meristems, lower plants have an apical cell.

aplanospore: a non-motile, asexual spore.

apocarpous: having a gynaecium with carpels free from each other.

apomixis: a form of a non-sexual reproduction found in some flowering plants, where meiosis does not occur in the ovule, and a diploid cell of the nucellus develops into the embryo, with apparently normal fruit and seed formation, *a.* apomictic.

appressorium: an enlargement of the germ tube of a parasitic fungus, that attaches to the host, and from which an infection peg grows.

archegoniophore: a stalk bearing archegonia or female receptacles.

archegonium: a typically flask-shaped female sex organ, made of a narrow neck formed of neck cells, containing neck canal cells, and a swollen venter containing the ovum; *pl.* archegonia.

areolae: small areas into which a thallus or some other structure is divided.

aril: an outgrowth on a seed; *a.* arillate.

ascocarp: a fruit body containing asci.

ascospore: one of the haploid spores produced inside an ascus by meiosis.

ascus: a sporangium containing ascospores; *pl.* asci.

asexual reproduction: another name for non-sexual reproduction.

auricles: in some grasses, two outgrowths from the leaf at the position of the ligule, which grasp the stem.

autonomic movement: a movement of a part of a plant in response to an internal stimulus, and which needs no external stimulus to occur.

autospore: a daughter cell formed inside an algal cell, and which has all the characteristics of its parent before being set free.

autotrophic: able to synthesise organic food material from inorganic substances by photosynthesis or chemosynthesis: *cf.* heterotrophic.

auxospore: in diatoms, a resting spore formed before the cell wall is made, and which restores the normal size of the organism.

axile placentation: the position of ovules in the ovary, where they lie on placentae in the middle of the ovary, in the angle formed by the meeting of septa.

basidiocarp: a fruit body bearing basidia.

basidiospore: one of (usually) four haploid spores produced by meiosis on the outside of a basidium.

basidium: the cell which bears basidiospores; *pl.* basidia.

berry: a succulent fruit, usually with many seeds, in which the whole pericarp becomes fleshy, e.g. tomato, gooseberry, sweet pepper, banana; see also hesperidium, pepo.

biennial: a plant which completes its life cycle in two years, passing the first winter usually in an underground storage organ; in the second winter the whole plant dies, and it passes this winter only as a seed in the ground, e.g. carrot, parsnip, beetroot; *cf.* annual, perennial.

bifid: forked deeply into two processes.

bilateral symmetry: having only one plane of symmetry, i.e. forming mirror images if cut in half along only one axis; in flowers also called zygomorphy.

biloproteins: in algae, blue and red pigments including phycocyanin and phycoerythrin, which contain a protein part complexed with a porphyrin; also called phycobilins and biliproteins.

binary fission: splitting into two.

biosynthesis: the use of the micro-organisms or enzymes derived from them to make (synthesise) a product required by man, e.g. using yeast to make alcohol.

biotechnology: the use of micro-organisms for biosynthesis, or for some other industrial or medicinal process.

body cell: one of the two cells into which the generative cell divides in a pollen grain (the other being the stalk cell) and which divides to form male gametes; also called spermatogenous cell, and thought to be equivalent to, and sometimes called, an antheridium.

bract: a leaf-like structure subtending a group of sex organs or sporangia, an inflorescence or a flower.

bracteole: a small bract on a flower stalk.

bran: the husk of a cereal grain, made of fused pericarp and testa.

153

budding: the production of daughter cells as outgrowths of a parent cell.

bulb: an organ of vegetative reproduction and perennation where food is stored in the bases of foliage leaves or in scale leaves surrounding one or more buds; the stem is very reduced and bears adventitious roots.

bulbil: a very small bulb, usually borne above ground, e.g. replacing a flower.

calyptra: the remains of an archegonium in bryophytes after fertilisation, forming the cap on a moss plant, and surrounding the young capsule before elongation of the seta in liverworts.

calyx: the collective name for the sepals.

cambium: a layer of meristematic cells in a position other than at the tip of a stem or root, capable of dividing to form secondary tissue such as wood.

capitulum: the inflorescence in the Compositae, made of very small flowers called florets, and surrounded by an involucre of bracts.

capsule: (1) in bryophytes, the structure on the sporophyte which contains the spores; (2) in angiosperms, a dry, many-seeded, dehiscent fruit made of two or more fused carpels, e.g. poppy, antirrhinum, bluebell.

carcerulus: a fruit which splits into one-seeded nutlets, as in the Labiatae where four nutlets are formed.

carotenes: a group of unsaturated hydrocarbons which are yellow, orange, or red.

carotenoids: a group of plant pigments which includes the carotenes and xanthophylls; also spelt carotinoids.

carpel: one of the units of which the gynaecium of an angiosperm is made up, and which is equivalent to the megasporophyll of pteridophytes and gymnosperms.

carpellate flower: a female flower.

caryopsis: the achene-like fruit in Gramineae, which has the testa and pericarp fused.

catkin: a pendulous inflorescence of reduced, sessile, usually unisexual flowers.

cell membrane: (1) the membrane, made of phospholipid and protein, which surrounds the cytoplasm of a cell; also called plasmalemma or plasma membrane; (2) any other membrane of the same structure in the cell, such as the endoplasmic reticulum or nuclear membrane.

cellulose: a polysaccharide made of glucose units, found in cell walls.

cell wall: a non-living layer surrounding the cell membrane, made of cellulose, hemicellulose, chitin, or other substances, and sometimes with additional material such as lignin or cutin.

chalaza: the end of an ovule away from the micropyle.

chemotaxis: the swimming of a free-living part of a plant, e.g. a spermatozoid, in relation to the direction of a chemical stimulus.

chitin: a mucopolysaccharide made of units of the amino sugar acetyl-glucosamine, found in the cell walls of many fungi.

chrysolaminarin: a polysaccharide food store in some algae; also called leucosin.

clamp connection: a hypha running across a septum in some Basidiomycetes.

class: a term used in classification for a group of orders which are thought to be related and to have a common ancestor.

coenocytic: filamentous but without septa (cross walls); also called aseptate.

coleoptile: a sheath covering the plumule.

coleorhiza: a sheath covering the radicle.

colonial: living as a group of similar cells forming a cluster, usually with little or no division of labour.

companion cell: in angiosperms, a cell with a nucleus which is associated with a sieve tube element and derived from the same parent cell.

conceptacle: a flask-shaped depression in the thallus of some algae, which contains hairs (paraphyses) for the absorption of minerals and secretion of mucilage and/or sex organs, and which opens by a pore.

cone: an arrangement of sporangiophores, microsporophylls and/or megasporophylls on the same axis, forming a conical structure; also called strobilus.

conidiophore: a hypha bearing conidia.

conidium: a non-sexual spore in some fungi, usually produced in chains; *pl.* conidia.

conjugation: the fusion of morphologically similar gametangia.

corm: an organ for vegetative reproduction and perennation, consisting of a short swollen upright stem surrounded by scale leaves, and containing one or more buds, e.g. crocus, gladiolus.

corolla: the collective name for the petals.

cotyledon: one of the first leaves of the plant, present in the seed as part of the embryo, and either remaining within the testa or emerging above ground and becoming green, forming one of the first foliage leaves.

cross fertilisation: the fusion of male and female gametes from genetically dissimilar plants, thus encouraging variation upon which ultimately the evolution of the species depends.

crossing: cross pollination or cross fertilisation.

cross pollination: the transfer of pollen from the anther to the stigma of another flower of the same species.

crozier: the crook-shaped young frond of a fern.

cuticle: a secondary cell wall of cutin, laid down on the epidermis of stems and leaves of land plants, which is impermeable to water and therefore aids water conservation.

cyclic: arranged in rings (whorls).

cypsela: the achene-like fruit in Compositae, which has the testa and pericarp separate, is made from an inferior ovary of two fused carpels, and often has a pappus of hairs (made from a modified calyx) for dispersal.

definitive nucleus: another name for secondary nucleus.

dehiscence: the shedding of spores or seeds, etc.

dehiscent: having fruit which stays on the plant and opens to release seeds; *cf.* indehiscent.

dichogamous: having stamens and stigmas maturing at different times; also called heterogamous; see also protandrous, protogynous; *n.* dichogamy; *cf.* homogamous.

dichotomous branching: branching produced by an apical cell dividing into two, each giving rise to a branch which then divides in the same way.

didynamous: having two long and two short stamens, as in the Labiatae.

dikaryotic mycelium: a mycelium having cells with two nuclei, one of each mating type; *cf.* monokaryotic mycelium.

dioecious: having male and female sex organs on different plants; *cf.* monoecious, hermaphrodite.

diploid: having two sets of chromosomes, twice the haploid number; symbol 2n.

diplontic: of a type of life cycle in algae where the plant body is diploid and only the gametes are haploid, e.g. *Fucus; cf.* haplontic, alternation of generations.

disc: the fleshy portion of the receptacle which sometimes secretes nectar.

division: the term used in classification for the main groups of plants. The nature of the divisions varies in different systems of classification, but members of the same division are thought to be related and to have had a common ancestor at an early stage of their evolutionary history; also called phylum.

dolipore septum: in the hyphae of some Basidiomycetes, septa which have pores that allow movement of food and cytoplasm along the hyphae.

dormant: viable but not germinating.

double fertilisation: in angiosperms, the condition of having two male gametes in the pollen tube, one of which fuses with the female gamete to form a zygote, the other of which fuses with the polar nuclei or with the secondary nucleus to form a triploid nucleus from which the endosperm develops.

drop mechanism: in gymnosperms, a system for catching pollen grains, consisting of a drop of liquid secreted by the micropyle.

drupe: a usually fleshy, often one-seeded fruit in which the pericarp is divided into an epicarp (skin), mesocarp (flesh) and endocarp (stone) which surrounds the seed, e.g. cherry, plum, apricot, etc.; also called stone fruit.

drupel (drupelet): a fruit made of many tiny drupes, e.g. blackberry.

egg apparatus: the synergids and ovum in an embryo sac.

egg cell: another name for ovum.

elaters: in liverworts, elongated, spirally thickened cells associated with spores and important in spore dispersal; similar structures, which are part of the cell wall, are found in horsetails where they are also called haptera.

elaterophore: in liverworts, a mass of elater-like cells attached to the base of the capsule, which aid spore dispersal.

embryo: the multicellular young plant developed from the zygote.

embryo sac: (1) the female gametophyte in spermatophytes; (2) another name for megaspore in spermatophytes, i.e. the female gametophyte at the unicellular stage.

endocarp: in fleshy fruits, the inner part of the pericarp, in drupes forming the stony layer; *cf.* epicarp, mesocarp.

endosperm: nutritive tissue in the seeds of gymnosperms and angiosperms, made from haploid female prothallus in gymnosperms, and from triploid tissue derived from the triple fusion nucleus in angiosperms.

endospermic seed: a seed in which the endosperm is not absorbed into the cotyledons; *cf.* non-endospermic seed.

entomophilous: having insect pollination; *n.* entomophily; *cf.* anemophilous.

epicalyx: a calyx-like structure below but near to the calyx.

epicarp: the outer part of the pericarp in fleshy fruits, usually the skin; also called exocarp; *cf.* endocarp, mesocarp.

epicotyl: the region between the base of the plumule and the cotyledons; *cf.* hypocotyl.

epidermis: the outer region of the plant body, small usually a single layer of cells.

epigeal germination: germination in which the cotyledons are brought above the soil; *cf.* hypogeal germination.

epigynous flower: a flower having sepals, petals and stamens inserted above the ovary, giving an inferior ovary; *cf.* perigynous, hypogynous.

epipetalous: attached to the petals.

essential organs: the androecium and gynaecium of a flower.

euglenoid movement: contraction of the cytoplasm within the pellicle, producing wriggling movements.

eukaryotic: having cells in which the DNA is organised into chromosomes, and is surrounded by a nuclear membrane; includes all organisms except bacteria and blue-green algae; *n.* eukaryote; *cf.* prokaryotic.

exine: the outer part of the wall of a pollen grain; *cf.* intine.

exocarp: another name for epicarp.

extrorse: having anthers dehiscing outwards; *cf.* introrse.

eyespot: in unicellular algae, a light-sensitive, pigmented region; also called stigma.

false fruit: a fruit which includes material of a structure other than the pericarp, e.g. hip, pome; also called pseudocarp.

family: a term used in classification for a group of genera thought to be closely related and to be descended from a common ancestor.

female gametophyte: a gametophyte derived from a megaspore.

fermentation: anaerobic respiration, particularly in micro-organisms.

fertilisation: the fusion of gametes to form a zygote in sexual reproduction; also called syngamy.

filament: (1) in general, a line of cells joined together; (2) in blue-green algae, a trichome surrounded by a mucilaginous sheath; (3) in flowering plants, the stalk of a stamen; *a.* filamentous.

fission: splitting into two or more parts.

flagellum: a whip-like extension of cytoplasm with which some cells can swim by lashing movements; *pl.* flagella.

floret: (1) a reduced flower in the inflorescence of the Compositae; it may be **tubular,** where all the petals are the same size forming a tube, or **ligulate,** where one petal is larger than the others and is strap-shaped: in an inflorescence with two kinds of floret, the outer florets are called **ray** florets, and the inner ones are called **disc** florets; (2) a reduced flower in other families.

follicle: a dry, dehiscent, usually many-seeded fruit formed from a single carpel, of which there are more than one in a flower, dehiscing along one suture; *cf.* pod.

fragmentation: reproduction by the breaking of (usually) a filament into parts, each of which develops into a new individual.

frond: (1) the flattened part a of thallus of a seaweed; (2) the leaf of a fern.

fruit: in angiosperms, a structure enclosing the seeds, made from the ovary after fertilisation, and sometimes also from associated structures.

fruit body: a multicellular structure bearing reproductive bodies, particularly spores; also called fructification, sporocarp, sporophore.

frustule: the cell wall of a diatom, with or without its living contents; the wall is shaped like a lidded box, with the larger part, the lid, called the epitheca, and the smaller part, the box, called the hypotheca.

funicle: an ovule stalk.

gametangium: a structure which produces gametes; *pl.* gametangia.

gamete: a haploid cell, two of which fuse in sexual reproduction to form a zygote; also called sex cell.

gametophyte: in a plant life cycle with alternation of generations, the haploid generation which bears the gametes.

gemma: a small multicellular bud in bryophytes, borne on the thallus in a gemma cup, or on a rhizoid; *pl.* gemmae.

genetic recombination: a parasexual process found in some blue-green algae, in which fusion and meiosis do not occur, but recombination of genes is achieved.

genus: a term used in classification for a group of closely related species; members of the same genus can sometimes breed together but usually do not form fertile offspring; *pl.* genera; *a.* generic.

geological time scale: the divisions of the ages of geological history, which are shown below.

geological time scale

Era	Period	Epoch
Cainozoic (Age of Mammals)	Quaternary	{ Recent { Pleistocene (Ice Age)
	Tertiary	⌠ Pliocene ⎪ Miocene ⎨ Oligocene ⎪ Eocene ⌡ Palaeocene
Mesozoic (Age of Reptiles)	Cretaceous (chalk formed) Jurassic Triassic	
Palaeozoic	Permian Carboniferous (Coal Measures) Devonian (Age of Fishes) Silurian Ordovician Cambrian	
Precambrian	Precambrian	

geotropism: a growth movement in response to the direction of gravity.

"germ": the embryo of a cereal grain.

germination: (1) all the stages in the development of a seed from the time of absorption of water through the micropyle to the opening of its leaves or cotyledons and the beginning of photosynthesis; (2) the early stages of growth of a spore, zygote, or pollen grain.

germling: a filament produced from a spore, which develops into a plant body; the term is usually used in algae, and the term germ tube is used in fungi.

germ tube: a hypha produced from a fungal spore; *cf.* germling.

gill: in fungi, a lamella-like structure on which basidiospores are formed.

glume: one of the bracts around the spikelet of a grass.

gluten: a protein found in some cereals, particularly wheat, which is capable of stretching and is essential for making leavened bread, since it enables bubbles of gas to become trapped in the dough.

glyoxalate cycle: a biochemical cycle in which fats are converted to sugars.

gonidia: the algal cells in a lichen; *sing.* gonidium.

gynaecium: the female part of the flower, consisting of one or more stigmas, styles and ovaries, and of one or more carpels which may be fused or free; also spelt gynoecium.

haploid: having one set of chromosomes; symbol n.
haplontic: of a type of life cycle in algae, where the plant body is haploid and only the zygote is diploid and immediately undergoes meiosis, e.g. *Spirogyra, Chlamydomonas; cf.* diplontic, alternation of generations.
hapteron: (1) an elater-like structure in horsetails; (2) another name for holdfast in algae; *pl.* haptera.
haptotropism: a growth movement in response to the direction of the stimulus of touch; also called thigmotropism.
haustorium: a hypha which secretes digestive enzymes and/or is used to absorb nutrients from the host; *pl.* haustoria.
herbaceous: non-woody, soft and green.
hermaphrodite: having male and female sex organs in the same flower or other reproductive structure; *cf.* dioecious, monoecious.
hesperidium: the citrus fruit, i.e. orange, lemon, grapefruit, etc., which is a modified berry in which the endocarp contains juice in hairs, the mesocarp forms the white ''zest'' and the epicarp forms the skin.
heterocyst: in blue-green algae, an enlarged, thick-walled vegetative cell which is thought to be concerned with nitrogen fixation, and may also be the point at which the filament breaks when hormogonia are formed.
heterogamous: another name for dichogamous; *n.* heterogamy.
heteromorphic alternation of generations: having the sporophyte and gametophyte different in appearance, e.g. *Laminaria* and all bryophytes, pteridophytes and spermatophytes; *cf.* isomorphic alternation of generations.
heterosporous: producing spores of different sizes; *n.* heterospory; *cf.* homosporous.
heterothallic: having physiologically different thalli is a species; mating can only occur between different types; *cf.* homothallic.
heterotrophic: obtaining food by the digestion of complex organic materials and including holozoic, parasitic and saprophytic nutrition; *cf.* autotrophic.
hilum: the scar on a seed coat, where it was attached to its stalk.
hip: a false fruit in which the receptacle becomes fleshy and surrounds one or more achenes.
holdfast: an organ which attaches a plant, particularly an alga, to a surface; also called hapteron.
holozoic: feeding like an animal, by engulfing food material and usually digesting it internally; sometimes used to mean heterotrophic.
homogamous: having stamens and stigmas maturing at the same time; *n.* homogamy; *cf.* dichogamous.
homologous: having the same ancestral origin.
homosporous: producing spores all of which are about the same size; *n.* homospory; *cf.* heterosporous.
homothallic: having only one type of thallus in a species; mating can occur between identical individuals or on the same individual.
hormogonia: in blue-green algae, short lengths of filament concerned with reproduction, which can move with a gliding motion; *sing.* hormogonium.
hygroscopic: taking up or losing water, usually resulting in movement.
hymenium: in a fruit body, the layer of cells (asci or basidia) which produce the spores (ascospores or basidiospores).
hypha: a filament of the plant body (mycelium) of a fungus; *pl.* hyphae; *a.* hyphal.
hypocotyl: the region between the base of the radicle and the cotyledons; it is the region where the arrangement of vascular tissue changes from that of a root to that of a stem; *cf.* epicotyl.

hypogeal germination: germination in which the cotyledons remain below the soil; *cf.* epigeal germination.
hypogynous flower: a flower having sepals, petals and stamens inserted below the ovary, giving a superior ovary; *cf.* epigynous, perigynous.

inbreeding: mating with an identical or genetically similar individual, thus reducing variation in the species; *cf.* outbreeding.
incompatibility: (1) the mechanism by which the pollen of one flower does not grow on the stigma of its own or certain other flowers; (2) the mechanism by which two identical strains of a lower plant do not mate; (3) in general, a mechanism which prevents self fertilisation.
indehiscent: having fruit which does not open to release the seeds, but the whole fruit is dispersed; *cf.* indehiscent.
indusium: in ferns, a structure covering the sorus; *pl.* indusia.
infection peg: in parasitic fungi, a structure produced from the appressorium, which penetrates the cuticle of the host, either by giving out enzymes or by mechanically boring through the host cell walls.
inferior ovary: an ovary which is sunken and fused with the receptacle, and has the perianth inserted around the top.
inflorescence: (1) all the flowers on one branch or stem; (2) in bryophytes, the antheridial or archegonial branches with their surrounding perichaetia.
integument: the coat of an ovule, which develops into the seed coat (testa) after fertilisation.
intercalary meristem: a meristem producing growth in length of the plant, formed somewhere other than at the tip.
internode: the length of stem between two nodes.
intine: the inner part of the wall of a pollen grain; *cf.* exine.
introrse: having anthers dehiscing inwards; *cf.* extrorse.
involucre: (1) in spermatophytes, a cluster of bracts forming a calyx-like structure; (2) in bryophytes, an outgrowth of the gametophyte forming a cup around the archegonium.
isogamy: the fusion of more-or-less identical gametes which are called isogametes; *a.* isogamous; *cf.* anisogamy, oogamy.
isomorphic alternation of generations: having the sporophyte and gametophyte the same in appearance. e.g. *Ulva; cf.* heteromorphic alternation of generations.

key fruit: another name for samara.

lamina: a blade, such as of a leaf or seaweed frond.
laminarin: a polysaccharide food store in some algae.
legume: (1) another name for pod; (2) a member of the Leguminosae.
lemma: the outer of the two bracts of a grass flower; *cf.* pale.
ligule: (1) in club mosses, a small scale borne near the base of the sporophyll; (2) in grasses, a membranous outgrowth of the leaf where the blade comes away from the stem; (3) in ligulate florets, the strap-shaped corolla.
locule: a cavity or compartment; *a.* locular.
lodicules: in grasses, the two perianth segments which are small, scale-like structures pressed close to the ovary.

male gametophyte: a gametophyte developed from a microspore.
medulla: the central region in the frond or stipe of a seaweed.

megagametophyte: another name for female gametophyte.

megaphyllous: having large leaves; also called macrophyllous.

megasporangium: a sporangium in which meiosis occurs to produce megaspores in pteridophytes; it corresponds roughly to the whole ovule, or more precisely to the nucellus of the ovule in spermatophytes; *pl.* megasporangia.

megaspore: a large spore produced by meiosis, which contains the female gametophyte; in spermatophytes may also be called the embryo sac.

megasporophyll: a sporophyll which bears megasporangia; in gymnosperms sometimes called an ovuliferous scale and in angiosperms called a carpel.

meiosis: nuclear or cell division which results in halving the chromosome number and produces four haploid, non-identical cells.

meristem: a group of cells which undergo mitosis and produce growth.

mesocarp: the middle part of the pericarp, usually the fleshy part, especially in drupes.

microgametophyte: another name for male gametophyte.

micropyle: the pore in the integument and testa.

microsporangium: a sporangium in which meiosis occurs to produce microspores; in spermatophytes also called pollen sac; *pl.* microsporangia.

microspore: a small spore produced by meiosis, containing the male gametophyte generation; in spermatophytes also called pollen grain .

microsporophyll: a sporophyll which bears microsporangia; in angiosperms modified to form a stamen.

mitosis: nuclear or cell division which results in the chromosome number remaining the same, and which produces two identical cells.

monoecious: having male and female flowers, or male and female sex organs, on the same plant but in different reproductive structures; *cf.* dioecious, hermaphrodite.

monokaryotic mycelium: a mycelium having cells each with one nucleus, which will later fuse with a mycelium of another mating type to form a dikaryotic mycelium.

mother cell: (1) a cell which undergoes meiosis to give haploid spores, e.g. pollen mother cell, megaspore mother cell, spore mother cell; (2) a cell that divides to produce other differentiated cells.

mould: the common name for a saprophytic fungus which causes decay.

mucilages: substances produced by plants which swell in water to make slimy solutions, and are chemically heterosaccharide derivatives.

mycelium: the mass of hyphae which make up the plant body in many fungi.

mycorrhiza: an association between a fungal mycelium and a plant root.

n: symbol for haploid; **2n:** symbol for diploid.

natural group: a group of plants which share many characteristics and which are thought to have one common ancestor, e.g. the flower families Cruciferae and Labiatae.

neck canal cells: cells which fill the neck of an archegonium and are extruded with the ventral canal cell to provide a chemical stimulus to attract spermatozoids.

neck cells: cells which make up the neck of the archegonium.

nectary: a structure which secretes nectar.

node: a position on a stem where a leaf or leaves arise.

non-endospermic seed: a seed in which the endosperm is absorbed into the cotyledons; *cf.* endospermic seed.

non-sexual reproduction: reproduction which does not involve the fusion of gametes; also called asexual reproduction.

nucellus: the tissue in the ovule where meiosis occurs, equivalent to the megasporangium of pteridophytes, although more loosely the whole ovule may be considered equivalent to the megasporangium.

nut: a dry, indehiscent, usually single-seeded fruit with a hard woody pericarp, often surrounded by a cupule made of bracts, e.g. hazel, chestnut, acorn.

nutlet: a one-seeded part of a fruit which breaks off at maturity.

offset: a very short runner, or the plant produced from it.

oogamy: the fusion of a small, motile male gamete with a large, non-motile female gamete; such gametes are called oogametes; *a*. oogamous; *cf.* anisogamy, isogamy.

oogonium: in some algae and fungi, the female sex organ; *pl.* oogonia.

oosphere: another name for ovum, especially in lower plants with oogamy.

oospore: a zygote formed by oogamy in lower plants where the ovum is called an oosphere; loosely another name for zygote; more correctly a zygote which has secreted a wall around itself.

operculum: a lid-like structure.

order: a term used in classification for a group of families which are thought to have been descended from a common ancestor.

ostiole: a pore.

outbreeding: mating with a genetically non-identical individual, thus promoting variation in the species; *cf.* inbreeding.

ovary: the part of the gynaecium which encloses the ovules, and which becomes the fruit after fertilisation.

ovule: in spermatophytes, a structure containing the embryo sac and nucellus and with coats called integuments, which after fertilisation develops into the seed; it is roughly equivalent to the megasporangium of pteridophytes, but the integuments are new structures in spermatophytes, and more precisely the nucellus of the ovule is equivalent to the megasporangium, since it is here that meiosis occurs to produce the megaspore (embryo sac).

ovuliferous scale: in gymnosperms, a structure bearing ovules, equivalent to a megasporophyll, and usually arranged in female cones.

ovum: a female gamete; also called egg cell, and in lower plants with oogamy also called oosphere.

pale (palea): the inner of the two bracts of a grass flower; *cf.* lemma.

palmella: a reproductive stage in some algae in which, after cell division of a parent cell, the daughter cells remain together and are surrounded by mucilage; also called palmelloid stage.

paramylum: a polysaccharide food store in some algae.

paraphyses: in lower plants, thread-like structures sometimes associated with reproductive cells or organs; *sing.* paraphysis.

parasitic: obtaining food from another living organism, the host, to which the parasite is attached, at least while it is feeding, and harming the host to some extent.

parenchymatous: in algae, having a plant body made of plate-like masses of similar cells, e.g. *Ulva*; the term is also used to mean thalloid.

parietal placentation: the position of ovules in an ovary where the ovules arise from the edge of the ovary walls or from a protrusion from them.

pedicel: a flower stalk.

peduncle: an inflorescence stalk.

pellicle: a thin outer covering, but not a cell wall.

peltate: describing a flat organ with a stalk inserted in the middle of the under surface, not at the edge, like a nasturtium leaf.

pendulous: hanging downwards.

pepo: a berry-like fruit with a hard rind, formed from an inferior ovary, found in members of the Cucurbitaceae, e.g. cucumber, marrow.

peptidoglycan: a macromolecule found only in the cell wall of prokaryotes, consisting of two amino sugars, N-acetyl-glucosamine and N-acetyl muramic acid, and an amino acid or peptide made of few amino acids; also called mucopeptide or murein.

perennation: a method of passing the winter, so that the plant survives from year to year.

perennial: a plant which survives vegetatively as well as by seed from year to year; herbaceous perennials die down above ground and pass the winter in underground storage organs; woody perennials stand the winter as trees or shrubs.

perianth: (1) in angiosperms, petals and sepals collectively; also called floral leaves or tepals; (2) in bryophytes, a somewhat tubular structure around the archegonium and sometimes the antheridium.

pericarp: the fruit wall, which is developed from the ovary wall.

perichaetia: enlarged leaf-like structures forming male and female rosettes in mosses; *sing.* perichaetium.

perigynous flower: a flower having a gynaecium on a concave receptacle with the perianth and androecium on its edge; intermediate in appearance between hypogynous and epigynous.

periplast: a thin outer covering, but not a cell wall.

peristome: the ring of teeth surrounding the mouth of a moss capsule, used in spore dispersal.

petal: one of the inner series of perianth lobes, if different from the outer, and particularly if other than green in colour.

petaloid: petal-like.

petiole: a leaf stalk.

phloem: tissue in vascular plants consisting of sieve tubes and (in angiosperms only) companion cells, and conducting mainly sugars.

photoreceptor: a structure sensitive to light.

phototaxis: a swimming movement of a free swimming plant or part of a plant towards (positive) or away from (negative) or at right angles to (dia-) the direction of a light source.

phototropism: a growth movement of part of a plant towards (positive) or away from (negative) or at right angles to (dia-) the direction of a light source.

phyllids: in bryophytes, leaf-like structures which are not true leaves because they do not have stomata or vascular tissue, and usually lack a cuticle.

phylum: another name for division.

phytoplankton: the plant part of the plankton.

pileus: the cap of a mushroom or toadstool.

pinna: a leaflet of a pinnate leaf; *pl.* pinnae.

pinnate leaf: a leaf made of more than three leaflets lying in two rows along a common stalk.

pinnule: a lobe of a pinna or sometimes of a pinnate leaf.

placenta: (1) the place where an ovule arises in an ovary; (2) the place where a sorus of sporangia arises in a fern; *pl.* placentae.

placentation: the position of placentae in the ovary.

plankton: the floating community of plants (phytoplankton) and animals (zooplankton) found near the surface of ponds, lakes, oceans, etc.

plasmodium: in slime moulds, a mass of amoeboid cells, forming the organism's body.

plumule: embryonic shoot.

pod: a dry, dehiscent fruit formed from a single carpel, containing many seeds, and usually dehiscing along both dorsal and ventral sutures; only one pod is present in a flower; also called legume; *cf.* follicle.

polar nuclei: in angiosperms, the two haploid nuclei in the middle of the embryo sac, which fuse with each other and with a male gamete to form the triploid triple fusion nucleus; in some cases the polar nuclei fuse together first to form the secondary nucleus.

pollen grain: a microspore in spermatophytes.

pollen sacs: the microsporangia in spermatophytes, i.e. the structures in which meiosis occurs to form pollen grains (microspores).

pollen tube: the structure which grows out of a pollen grain in the direction of the ovule, and carries the male gametes to the embryo sac.

pollination: the transfer of pollen from an anther to a stigma or from a male cone to a female cone.

polypetalous: having free petals; *cf.* sympetalous.

polysepalous: having free sepals; *cf.* symsepalous.

pome: the fleshy false fruit found in apples, pears, and some other Rosaceae, in which the receptacle surrounds the ovary and is fused to it; the receptacle becomes the fleshy part of the fruit that is eaten, while the pericarp (developed from the ovary wall) becomes the core.

primary endosperm nucleus: another name for triple fusion nucleus.

primary growth: growth from the apical meristem, not from the cambium, and producing soft, herbaceous tissue.

prokaryotic: having cells in which the DNA lies in a particular region of the cell, but is not organised into chromosomes or surrounded by a nuclear membrane; includes bacteria and blue-green algae only; *n.* prokaryote; *cf.* eukaryotic.

prop root: a root arising from the stem above ground and growing down to the soil, thus supporting the stem.

prostrate: creeping; growing close to the ground.

protandrous: having the stamens maturing before the stigmas, encouraging cross pollination. *n.* protandry; *cf.* homogamous, protogynous.

prothallus: the structure forming the gametophyte generation in pteridophytes and gymnosperms: *pl.* prothalli.

protogynous: having the stigmas maturing before the stamens, encouraging cross pollination; *n.* protogyny; *cf.* homogamous, protandrous.

protonema: in some bryophytes, a filamentous stage produced from a spore, and from which the gametophyte plant arises.

protoplast: a unit consisting of a nucleus, cytoplasm and cell membrane, which may or may not be surrounded by a cell wall.

pseudocarp: (1) another name for false fruit; (2) a particular kind of false fruit, having achenes embedded in the outer surface of a fleshy receptacle, e.g. strawberry.

pseudomycelium: a structure formed by buds on a unicellular fungus such as yeast, where the buds remain attached to the parent, forming a body resembling a mycelium.

pseudoparenchyma: a mass of closely interwoven hyphae or filaments, which appears like parenchyma in section; *a.* pseudoparenchymatous.

pulse: a collective name for the edible seeds of the family Leguminosae,

e.g. peas, beans, lentils.
pyrenoid: in a chloroplast, a small protein body around which starch or paramylum is formed.

raceme: an inflorescence with stalked flowers, having the oldest flowers at the base, and the youngest flowers and the growing point at the top; *a.* racemose.
radial symmetry: having many planes of symmetry, i.e. forming mirror image halves if cut along any diameter; in flowers also called actinomorphic.
radicle: embryonic root.
ramenta: the golden-brown scale leaves covering young fern fronds and which are retained on the frond petiole.
raphe: in the frustule of some diatoms, a slit running along the longitudinal axis from one polar nodule to the other.
receptacle: (1) in some algae, the swollen tips of the thallus in which conceptacles bearing sex organs are housed; (2) in some liverworts, the umbrella-like structures on which antheridia and archegonia are borne; (3) in angiosperms, the top of the flower stalk from which the floral parts arise.
reflexed: bent backwards, e.g. sepals in bulbous buttercup.
rhizoid: a unicellular or filamentous thread-like structure, often acting as a root-like organ; *a.* rhizoidal.
rhizome: a horizontal and usually underground stem; *a.* rhizomatous.
rhizomorph: in some fungi, a thick thread made by union of hyphae.
rhizophore: in some club mosses, a structure arising from the stem and ending in a root.
root cap: a hollow cone of cells which covers the growing tip of a root and protects it from damage as it pushes through the soil.
root hair: a unicellular outgrowth of an epidermal (piliferous layer) cell of roots, used for the absorption of water and mineral salts.
rootstock: a short, upright, underground stem.
root tuber: a swollen root used for food storage, e.g. sweet potato, dahlia.
runner: a horizontal stem running along the surface of the ground, producing small plants at nodes between very long internodes, and used for vegetative reproduction but not for perennation.

samara: an achene-like fruit, of which part of the wall is flattened forming a wing, e.g. sycamore, maple, ash, elm; also called key fruit.
saprophytic: obtaining food from dead or organic material, the host, with which the saprophyte forms an intimate association, and causes the host to decay.
scutellum: the modified cotyledon in grasses.
secondary growth or thickening: growth produced by division of cells of the cambium and cork cambium, and including wood and bark.
secondary nucleus: in the embryo sac of some angiosperms, a diploid nucleus formed by the fusion of the two haploid polar nuclei; it later fuses with a male gamete to form a triploid nucleus which develops into the endosperm; also called definitive nucleus.
seed: in spermatophytes, a reproductive structure developed from a fertilised ovule, and consisting of an embryo, a food store (endosperm) and a coat (testa).
self fertilisation: the fusion of male and female gametes from the same plant, or from two genetically identical ones.
selfing: self pollination or self fertilisation.
self pollination: the transfer of pollen from the anther to the stigma of the same flower.

sepal: a member of the outer series of perianth lobes, if different from the inner, particularly when green.
sepaloid: sepal-like.
septate: having partitions (septa).
sessile: not stalked.
seta: in some bryophytes, the diploid stalk on which the capsule is borne, and which forms part of the sporophyte.
sex organ: a structure in which gametes are formed; sometimes called gametangium.
sexual reproduction: reproduction involving the fusion of gametes to form a zygote.
silicula: in Cruciferae, a capsule-like fruit almost as wide as it is long, made of two fused carpels, with parietal placentation, but having two compartments made by a false septum.
siliqua: in Cruciferae, a fruit which is like a silicula, but is long and narrow.
somatic: a structure or process that is other than reproductive.
soredium: a small dispersive and reproductive body in lichens, made of a few algal cells surrounded by some fungal hyphae; *pl.* soredia.
sorus: a group of sporangia; *pl.* sori.
species: a term used in classification for a group of plants which usually have many characteristics in common and can interbreed to form fertile offspring; a species has two names, the first being the generic name and the second the specific epithet.
sperm: abbreviation for spermatozoid.
spermatogenous cell: another name for body cell in gymnosperms.
spermatozoid: in many lower plants, a motile male gamete; abbreviated to sperm, also called antherozoid.
spike: a simple, long, racemose inflorescence of unstalked flowers.
spikelet: a unit of the inflorescence of a grass, consisting of two glumes (bracts) surrounding several flowers.
sporangiophore: a filament or hypha bearing one or more sporangia.
sporangium: a structure containing spores; *pl.* sporangia.
spore: a small unicellular or few-celled dispersive and usually reproductive body.
sporocarp: another name for a fruit body.
sporogonium: another name for the sporophyte generation in bryophytes.
sporophore: another name for a fruit body.
sporophyll: a leaf-like structure, or one that is thought to have evolved from a leaf, bearing or subtending sporangia; if it bears micro- or megasporangia it is called a micro- or megasporophyll.
sporophyte: in a plant with alternation of generations, the diploid generation which bears the spores.
stalk cell: one of the two cells into which the generative cell divides in a pollen grain, the other being the body cell; also called sterile cell.
stamen: in angiosperms, a structure evolved from a microsporophyll and consisting of an anther (containing pollen sacs) and a filament (stalk).
staminate flower: a male flower.
staminode: a sterile and often reduced stamen.
stem tuber: an underground swollen part of a stem at the end of a rhizome, storing food and bearing leaf scars and buds which can develop into new plants, e.g. potato, Jerusalem artichoke.
sterigma: an outgrowth of a basidium into which the nuclei of potential basidiospores migrate; *pl.* sterigmata.

sterile cell: another name for stalk cell.
stigma: (1) the receptive surface of a gynaecium on which pollen grains land; (2) in algae, another name for eyespot.
stipe: a stalk, particularly of the fruit body of a fungus or thallus of a seaweed.
stipule: a scale-like or leaf-like outgrowth of the stem, usually at the base of the petiole and sometimes attached to it.
stolon: a horizontal stem growing along the surface of the soil, producing young plants at its nodes, between short internodes, and used for vegetative reproduction but not for perennation.
stoma: a pore in the epidermis, which can be opened or closed by changes in shape of the guard cells which surround it; *pl.* stomata.
stomium: (1) the thin-walled cells in the wall of a fern sporangium where the wall splits at dehiscence; (2) the thin-walled part of an anther where dehiscence occurs.
strobilus: another name for cone.
style: the part of the gynaecium between the stigma and ovary.
subhymenium: the layer under the hymenium.
substrate: non-living material to which a plant is attached; also called substratum.
sucker: a short, underground stem which quickly comes to the surface and produces a new plant, e.g. roses, bananas.
superior ovary: an ovary which has the perianth inserted around its base.
swarmer: another name for zoospore.
symbiosis: an intimate partnership between two organisms of different species, called symbionts, which is for their mutual benefit; *a,* symbiotic.
sympetalous: having fused petals; *cf.* polypetalous.
symsepalous: having fused sepals; *cf.* polysepalous.
syncarpous: broadly, having carpels joined to one another; more strictly, having fused carpels separated by septa and axile placentation.
synergids: in the embryo sac of angiosperms, the haploid cells on either side of the ovum, the synergids and ovum together forming the egg apparatus.
syngamy: another name for fertilisation.

tap root: the main, primary root which develops from the radicle of the seed.
tegmen: the seed coat formed from the inner integument of the ovule; *cf.* testa.
testa: loosely, the seed coat, developed from the integuments of the ovule; more precisely, the coat formed from the outer integument of the ovule; *cf.* tegmen.
thallus: a multicellular plant body which is not differentiated into stem, root and leaves; *a.* thalloid.
thigmotropism: another name for haptotropism.
thylakoid: one of the membranes, usually of the grana, on which chlorophyll is situated.
tiller: a branch developed at a basal node in grasses.
torus: the receptacle, especially when it becomes fleshy, as in strawberry.
tracheid: a lignified water-conducting cell with no living contents.
trama: the central region of a gill, below the subhymenium.
trichome: in blue-green algae, a filament of cells which is not surrounded by a mucilaginous sheath.
trilocular: having three compartments.
triple fusion nucleus: in angiosperms, a triploid nucleus formed by the fusion of a male gamete with the polar nuclei or with the secondary nucleus, and which divides to form the endosperm; also called primary endosperm nucleus.
tube nucleus: a nucleus in a pollen grain, which migrates into the pollen tube, but disintegrates before fertilisation; also called vegetative nucleus.

tuber: a swollen underground food-storing organ; see stem tuber, root tuber.

umbel: an inflorescence in which the flower stalks arise from the same point at the top of the stem, often forming an umbrella-like shape.
unilocular: having a single compartment.
unisexual: having organs of one sex only.

vascular bundle: xylem and phloem found together in a strand, sometimes separated by cambium.
vascular plant: a plant having vascular tissue.
vascular tissue: conducting tissue made of xylem and phloem.
vegetative nucleus: another name for tube nucleus.
vegetative reproduction or propagation: non-sexual reproduction by multicellular propagules rather than spores.
venter: the lower, swollen part of an archegonium which contains the ovum.
vessel: a long, water-conducting tube of xylem, in which cross walls have broken down, the cell walls are lignified, and the cells are dead.
viable: being alive but dormant, not germinating.

whorl: more than two organs of the same kind at the same level, forming a ring.
wood: tissue composed of xylem, particularly secondary xylem.
woody: having secondary xylem.

xanthophylls: a group of yellow to brown plant pigments, which are oxygenated derivatives of carotenes.
xeromorphic: adapted to dry conditions.
xylem: in vascular plants, water-conducting tissue made of vessels and tracheids.
XY mechanism: (1) a mechanism determining sex in plants with alternation of generations where the sporophyte is 2n + XY and the gametophytes are n + X (female) and n + Y (male) produced in equal numbers; (2) a mechanism determining sex in dioecious diploid plants where the male contains the sex chromosomes XY and the female contains sex chromosomes XX, e.g. *Ginkgo*.

zoosporangium: a structure which contains zoospores; *pl.* zoosporangia.
zoospore: a motile, non-sexual reproductive cell, swimming by one or more flagella; also called swarm cell, swarmer.
zygomorphic: bilaterally symmetrical; *cf.* actinomorphic.
zygospore: a thick-walled resting spore formed by the fusion of undifferentiated gametangia, or by isogamy or anisogamy.
zygote: in sexual reproduction, a diploid cell formed by fusion of two gametes.

Index